FUNDAMENTALS OF COMPUTER LOGIC

THE ELLIS HORWOOD SERIES IN
COMPUTERS AND THEIR APPLICATIONS

Series Editor: BRIAN MEEK
Computer Unit, Queen Elizabeth College, University of London

The series aims to provide up-to-date and readable texts on the theory and practice of computing, with particular though not exclusive emphasis on computer applications. Preference is given in planning the series to new or developing areas, or to new approaches in established areas.

The books will usually be at the level of introductory or advanced undergraduate courses. In most cases they will be suitable as course texts, with their use in industrial and commercial fields always kept in mind. Together they will provide a valuable nucleus for a computing science library.

Published and in active publication

THE DARTMOUTH TIME SHARING SYSTEM
G. M. BULL, The Hatfield Polytechnic

THE MICROCHIP AS AN APPROPRIATE TECHNOLOGY
Dr. A. BURNS, The Computing Laboratory, Bradford University

INTERACTIVE COMPUTER GRAPHICS IN SCIENCE TEACHING
Edited by J. McKENZIE, University College, London, L. ELTON, University of Surrey, R. LEWIS, Chelsea College, London.

INTRODUCTORY ALGOL 68 PROGRAMMING
D. F. BRAILSFORD and A. N. WALKER, University of Nottingham.

GUIDE TO GOOD PROGRAMMING PRACTICE
Edited by B. L. MEEK, Queen Elizabeth College, London and P. HEATH, Plymouth Polytechnic.

DYNAMIC REGRESSION: Theory and Algorithms
L. J. SLATER, Department of Applied Engineering, Cambridge University and H. M. PESARAN, Trinity College, Cambridge.

CLUSTER ANALYSIS ALGORITHMS: For Data Reduction and Classification of Objects
H. SPATH, Professor of Mathematics, Oldenburg University.

FOUNDATIONS OF PROGRAMMING WITH PASCAL
LAWRIE MOORE, Birkbeck College, London.

RECURSIVE FUNCTIONS IN COMPUTER SCIENCE
R. PETER, formerly Eotvos Lorand University of Budapest.

SOFTWARE ENGINEERING
K. GEWALD, G. HAAKE and W. PFADLER, Siemens AG, Munich

PROGRAMMING LANGUAGE STANDARDISATION
Edited by B. L. MEEK, Queen Elizabeth College, London and I. D. HILL, Clinical Research Centre, Harrow.

FUNDAMENTALS OF COMPUTER LOGIC
D. HUTCHISON, University of Strathclyde.

SYSTEMS ANALYSIS AND DESIGN FOR COMPUTER APPLICATION
D. MILLINGTON, University of Strathclyde.

ADA: A PROGRAMMER'S CONVERSION COURSE
M. J. STRATFORD-COLLINS, U.S.A.

FUNDAMENTALS OF COMPUTER LOGIC

DAVID HUTCHISON, B.Sc., M.Tech.
Department of Computer Science
University of Strathclyde

ELLIS HORWOOD LIMITED
Publishers · Chichester

Halsted Press: a division of
JOHN WILEY & SONS
New York · Chichester · Brisbane · Toronto

First published in 1981 by

ELLIS HORWOOD LIMITED
Market Cross House, Cooper Street, Chichester, West Sussex, PO19 1EB, England

The publisher's colophon is reproduced from James Gillison's drawing of the ancient Market Cross, Chichester.

Distributors:

Australia, New Zealand, South-east Asia:
Jacaranda-Wiley Ltd., Jacaranda Press,
JOHN WILEY & SONS INC.,
G.P.O. Box 859, Brisbane, Queensland 40001, Australia

Canada:
JOHN WILEY & SONS CANADA LIMITED
22 Worcester Road, Rexdale, Ontario, Canada.

Europe, Africa:
JOHN WILEY & SONS LIMITED
Baffins Lane, Chichester, West Sussex, England.

North and South America and the rest of the world:
Halsted Press: a division of
JOHN WILEY & SONS
605 Third Avenue, New York, N.Y. 10016, U.S.A.

British Library Cataloguing in Publication Data
Hutchison, David
    Fundamentals of computer logic. —
    (Ellis Horwood series in computers and their applications).
    1. Electronic digital computers — Circuits
    2. Logic circuits
    I. Title
    621.3819'5835   TK7888.4   80–42028

ISBN 0–85312–258–X (Ellis Horwood Ltd., Publishers — Library Edn.)
ISBN 0–85312–305–5 (Ellis Horwood Ltd., Publishers — Student Edn.)
ISBN 0–470–27117–5 (Halsted Press)

Typeset in Press Roman by Ellis Horwood Ltd.
Printed in Great Britain by Butler and Tanner Ltd., Frome, Somerset.

Table of Contents

Preface 7

Chapter 1 The Structure of Computers
1.1   Introduction. 9
1.2   Computer logic 11
1.3   Structural layers 22

Chapter 2 Logic Building Blocks
2.1   Logic symbolism 24
2.2   Boolean algebra. 30
2.3   Logic families 35
2.4   Integrated circuit building blocks 44

Chapter 3 Combinational and Sequential Logic
3.1   Representation of logic circuits 57
3.2   Combinational logic design 65
3.3   Sequential logic design 90

Chapter 4 Logic Circuits in Practice
4.1   Design examples 110
4.2   Circuit problems in practice. 131
4.3   Analysis of logic circuits 140
4.4   Other approaches to logic implementation 143

Chapter 5 Computer Logic Design
5.1   Computer logic circuits. 150
5.2   Control structure. 164
5.3   Microprogramming 173

Chapter 6 The Hardware/Software Interface
    6.1   Interdependence of hardware and software. 178
    6.2   Summary. 185

Reading List. 187

Appendix. 193

Index 207

Author's Preface

This book describes the structure of computers through a study of their underlying logic circuits. The approach used is to present the material in a bottom-up way, starting with a description of the basic building blocks of logic circuits and proceeding in a series of layers to build a picture of how computers are designed and constructed. Although the material is principally concerned with computer hardware, some attention is paid to the mutual dependence of hardware and software requirements in computer design.

While the book is aimed primarily at first- and second-year undergraduate students at Universities and Polytechnics — in computer science, microprocessor studies and related engineering subjects — the approach is intended also to benefit those with an interest in computers from a mainly hardware point of view. Those who wish to learn about logic building blocks and their use in designing logic circuits, but not necessarily their application in computers, are also catered for since these topics — the core of the book — are essentially self-standing. A little background knowledge is required: a familiarity with binary numbers and an acquaintance with the notion of programming. The reader who has attended a short course on computer appreciation will be well prepared.

Chapter 1 sets the scene of the book by presenting a brief history of computers and outlining the structural layers of a modern computer. In Chapter 2 the building blocks of logic are introduced from both an abstract and a physical point of view: the theory of Boolean algebra and the implementation tools of integrated circuits ('chips') are brought together. Chapter 3 classifies logic circuits into the combinational and sequential varieties, and describes techniques for designing both types of circuit, with worked examples in each case. Further worked design examples are presented in Chapter 4, along with other aspects of the uses of logic circuits in practice. Several logic circuits used in computers are introduced before Chapter 5, which illustrates how the circuits fit together to implement the major functional units in a computer. Particular attention is paid to the design of arithmetic and control units. The last Chapter, 6, discusses the interdependence of hardware and software in computers and ends by commenting on the implications for computer design of new hardware technologies and advances in software techniques.

An Appendix contains a sample of typical literature available from a semiconductor manufacturer, describing some of the integrated circuits with which logic circuits, and computers, can be constructed. In the text the importance of referring to, and understanding, such data sheets is emphasised. The Reading List contains chapter-by-chapter recommendations for further reading. No references are included in the text but the annotations with each title in the List direct the reader to suitable sources for specific topics.

Students using this book as a course text will greatly benefit from a short practical course which illustrates the use of integrated circuits in logic design and implementation. Both the practical work and the choice of any design problems for students to tackle are best left to the discretion, and ingenuity, of the course lecturer.

Thanks are due to the second-year students of Computer Science at the University of Strathclyde who have helped, wittingly or otherwise, to evolve the approach used in this book by their participation in the course on Logic Design during the past three years. My thanks in general to my colleagues in the Department of Computer Science, and in particular to Miss Agnes Wisley and Mrs Margaret McDougall for their help in typing the manuscript. Brian Meek (the Series Editor) and Michael Horwood have helped give this book shape and direction and I am most grateful to them. Lastly, the book would never have been written without the help and support of two people, Ian Campbell and Ruth Hutchison.

Glasgow, August 1980

CHAPTER 1

The structure of computers

1.1 INTRODUCTION

Computers have two major ingredients: hardware and software. *Hardware* is the collective term used to describe the physical units of the computer — the processor, memory and peripheral devices, including all mechanical and electronic components. *Software,* on the other hand, refers in general to the programs which cause the computer hardware to obey specific sequences of instructions. The physical realisation of software is either a set of instructions, written in a particular language, on a piece of paper or a pattern of binary digits (*bits*) in the memory hardware of a computer.

Very often a distinction is made between system software and applications software. *System software,* sometimes referred to confusingly as simply software, is a set of programs written by the manufacturer or supplier of the hardware (or in some cases by the users themselves). This software consists of an *operating system* which controls the basic functions of the hardware, and a set of utility programs such as language translators, editors and debugging aids. Depending on the intended application area of the computer its operating system may provide more or less extensive facilities: in a large multi-access machine the operating system may have to service the widely-different needs of its many concurrent users and support a complex file system held on magnetic discs and tape. At the opposite extreme a small single-user computer may require an operating system which simply allows single programs to be loaded and started, and memory locations to be inspected and their contents altered. All computers require some system software which enables users to operate them easily. The extent to which utility programs are provided also depends on how the computer is used: systems on which programs are constantly being developed would typically have available a variety of high-level language compilers and a file editor, whereas dedicated systems in which the programs are fixed and proven may provide no such utilities along with the operating system.

Applications software, usually written by users or software houses, tailors a computer system (already provided with system software) to a particular task

or application. Typically applications software is written in a high-level programming language such as COBOL, FORTRAN or Pascal, and in this way the
hardware features of the computer are hidden from the programmer. The
compiler translates the user program into machine-code instructions, the form in
which the computer can understand the programmer's intentions. Interaction
with the outside world, via the peripheral devices, is handled on behalf of the
programmer by the operating system; the programmer simply indicates the need
for input or output by means of a high-level statement such as read(X) or
print(X) in the program text. In special circumstances, such as cases in which the
speed of operation of programs is critical, applications may be programmed in
assembly language, a symbolic form of the machine-code which computers
understand. The central part, or kernel, of an operating system is often written
in this machine-oriented form. Applications written in assembly language also
rely on the presence of an operating system to control the resources of the
computer hardware and to provide a machine-independent user interface.

In very general terms the structure of a computer system can be illustrated
as in Fig. 1.1. This shows the three components introduced above — hardware,
system software, and applications software — ordered in a hierarchy or a set of
layers. The computer hardware is the lowest, or most basic, of the layers, while

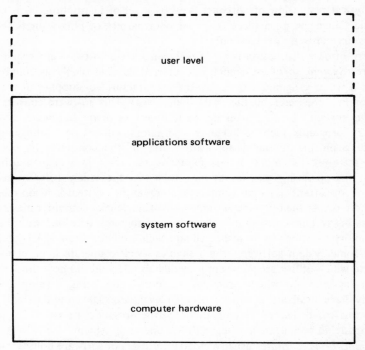

Fig. 1.1 — The broad layers of a computer system.

the highest layer in the computer hierarchy is the user level. Strictly speaking the user level is not a part of the computer as such; this is indicated by the dashed lines in the diagram.

Each layer provides a set of facilities for the one above. The architecture of the computer hardware determines the nature of the machine features which the system software has to handle; the type of use of the computer system influences the facilities which the system software offers to the applications programmer; and lastly, but very importantly, the applications level should provide a friendly set of facilities to the users. The boundary between one layer and the next is termed an *interface*: as we go from the hardware to the user level the interface facilities offered are more and more high-level, in other words further from the low-level machine features and somewhat closer to a form which people can easily understand. The *man-machine interface,* as the highest-level interface is called, is increasingly important with computer systems finding their way into every office and factory, where non-expert users are called upon to operate them.

*   *   *

The scope of this book falls well short of the man-machine interface, and short also of applications software. It deals with computer hardware, and to some extent with the interface between hardware and system software. More specifically the book is about *computer logic*, the electronic rather than the mechanical aspects of computer hardware — the logic circuits as opposed to the moving parts such as magnetic discs or line printers. The remainder of this chapter outlines the context in which the main material of the text is set.

1.2 COMPUTER LOGIC

To explain the meaning of computer logic let us look a little more deeply into the hardware layer of the computer system. It should be borne in mind that in this book we are dealing exclusively with *digital* as opposed to *analogue* computers. Analogue computers represent information in the form of continuously varying voltages or currents and are used in special-purpose applications like the design of automobile suspension systems where the physical system can be modelled or simulated by the computer.

A brief history

Digital computers represent information in discrete binary (two-valued) form and have evolved from the calculating machines of the last century, including Charles Babbage's design for an Analytical Engine (1837) and Hollerith's Electric Tabulating System (1889). The first electronic computer was the ENIAC (1946), inspired by a memorandum of J. W. Mauchly in 1942 within the Moore School of Electrical Engineering at the University of Pennsylvania. The ENIAC (Electronic Numerical Integrator and Calculator) took three years to build and was large-scale in every way. It contained some 19,000 valves, weighed 30 tons and consumed 200 kilowatts of electricity. It was also extremely fast by the standards of the day, being able to multiply two 10-digit decimal numbers in 3 milliseconds. However, the effort of programming the ENIAC was such as to discourage its use for any other than extensive computational problems, since it had to be programmed manually by plugging and unplugging sets of connecting wires. Data could be entered using a punched card reader, and results output on punched cards or on an electric typewriter. A large team was responsible for the design and construction of the ENIAC, most notably J. P. Eckert and J. W. Mauchly who in 1947 jointly founded a company to produce computers commercially. One of their first products was called the UNIVAC (Universal Autommatic Computer). Later their company became the UNIVAC division of the Sperry-Rand Corporation, which along with IBM began selling computers successfully in the early 1950s.

Another member of the Moore School team, John von Neumann (1903–1957), is credited with the idea now seen as the final step in the development of the general-purpose computer. This is the idea of a *stored-program* machine in which program and data share a common memory. The most important consequence is that programming is made very much easier; thus the computer possesses a generalised instruction set, fixed into its hardware, and the program — consisting of a sequence of appropriately chosen instructions — can be read in via a punched-card reader in the same way as the data. An additional consequence, one that has had less lasting significance, is that programs can be made to modify their own instructions.

There is evidence to suggest that others before von Neumann had the notion of a stored-program computer, notably Konrad Zuse (in his 1936 paper), who produced in Germany in the 1930s a series of mechanical and electro-mechanical computers called Z1, Z2 and Z3; A. M. Turing, whose 1936 abstract model of a computer — called a Turing machine — formed the basis for much of the present-day knowledge of the theory of computation; and even Charles Babbage, although in his case perhaps the suggestion is closer to speculation than in the others. However, it is certain that von Neumann's draft report on the EDVAC (Electronic Discrete Variable Computer) written in 1945 contains the earliest documented presentation of the stored-program idea. The EDVAC was the successor to ENIAC and contained several design changes which originated during the ENIAC project. The main differences were that it was a binary rather than a decimal machine and that it had a much larger memory: 1K (or 1024) 40-bit words of mercury delay-line main store, plus a secondary, slower magnetic store 20K words in size. This machine did not become operational until 1951.

Meantime a report written for the U.S. Army Ordnance Department in 1946 by Burks, Goldstine and von Neumann proposed a methodology for designing computers. This report was the first of a series which led to yet another machine called the IAS. In effect its proposals summarise the characteristics of *first-generation* digital computers. Burks, Goldstine and von Neumann suggest the following features:

— main units: control
                 arithmetic
                 memory
                 input/output communication
— program and data sharing the memory
— binary internal forms
— a synchronous clock system
— the use of subroutines
— possible adder hardware, but multiplier software
— the use of an accumulator register
— parallel operation for memory accessing
— diagnostic/single-step provision.

All these features are to be found in the majority of present-day computers. They characterise what has become known as the *von Neumann machine*. A schematic diagram of such a machine is shown in Fig. 1.2. This gives us an outline description of the typical hardware structure of a computer.

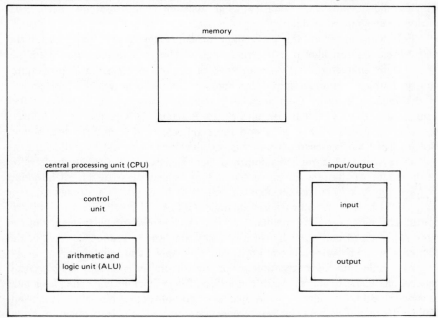

Fig. 1.2 — Outline computer structure.

There are three main parts:

(1) the memory, in which program and data co-reside.

(2) the central processing unit (CPU) which in turn has two components —
the arithmetic and logic unit (ALU) in which all calculations are performed, the results being held in the associated accumulator register;
the control unit, whose job it is to fetch and obey program instructions
from the memory, and to co-ordinate the activities of the other units.

(3) input and output units, which respectively read information into the
computer and print it out.

In the late 1940s and early 1950s many computers were built. In Britain
there was notable work at Manchester and at Cambridge. Probably the first
working stored-program computer was a small experimental machine built at
Manchester University in 1948 by F. C. Williams and T. Kilburn. The range of
machines produced at Manchester culminated (much later, around 1961) in the
famous ATLAS computer which had a *one-level store,* the forerunner of the
present-day virtual memory systems in which the (fast) main memory and the
(slow) magnetic backing store are seen by the programmer as effectively a single
large memory. At Cambridge M. V. Wilkes and others designed and built the
EDSAC (Electronic Delay Storage Automatic Calculator), a machine modelled
on the lines of the EDVAC. It was completed in 1949 and had a fixed program
which could nowadays be described as an assembler and loader, an early contri-
bution to programming aids.

Following on from the early days of computer design, from about the
mid 1950s, we can identify the emergence of the *second-generation* machine.
These were characterised by improvements in both hardware and programmability. Perhaps the most important hardware innovation was the replacement
of the vacuum tube by the *transistor,* a semiconductor device invented at Bell
Telephone Labs. by J. Bardeen and W. H. Brattain. This permitted computers
to be built which were smaller and more reliable, and consumed less power.
Not until the surface-barrier transistor was developed in 1954 by Philco did the
operating speed of computers improve significantly, however. Thereafter the
development of discrete transistor technology continued with the introduction
of *logic families* called direct-coupled transistor logic (DCTL), diode-transistor
logic (DTL), and resistor-transistor logic (RTL). These represented efforts
continually being made by the manufacturers to improve the performance of the
basic elements of computer hardware. A key aim was to reduce the cost of the
elements, while nevertheless also improving their speed and reliability.

Alongside the new technology of transistors in characterising second-
generation machines stands the introduction of *high-level programming languages*
as a major aid to speeding up the process of computer programming. The inten-
tion of the high-level language was to permit programming to be problem-oriented
and machine-independent. A *compiler* (or translator program) converts the high-

level language programs into machine-code specific to the computer which will run the program. FORTRAN (Formula Translation) was the first widely-used high-level programming language. It was developed by a group at IBM under the direction of John Backus between 1954 and 1957. COBOL (Common Business Oriented Language) followed in 1959, intended mainly for business applications, in contrast to Fortran which was designed specifically to be used in scientific work. ALGOL (Algorithmic Language), specified in 1960 and revised in 1962 by an international committee including, amongst others, John Backus and Peter Naur, was another important language designed during the second generation of computers. Other languages, now obsolete, were being designed and compilers implemented. The first system software was now beginning to be produced by computer manufacturers and was supplied as part of a package along with the computers themselves. This early system software tended to consist of compilers and rudimentary operating systems.

Apart from changes in technology and software, the architecture and logical design of computers were developing too. Second-generation machines tended to have a floating-point arithmetic unit; index registers and indirect addressing became standard; with magnetic core main memory the design of the CPU tended to be strongly influenced by the timing of memory accesses; and synchronous operations (that is linked to a common timing source) became very widely used.

Previously, asynchronous operations dominated: in this scheme the component parts of an operation (some slow, others fast) were allowed to proceed at their own pace, and job completion signals indicated that the next phase could begin. In computers of the second generation onwards the cycle of events within the machine was controlled by a central clock, both CPU and memory operations being synchronised from its timing pulses. The use of index registers was pioneered by the Manchester University team: these fast-access storage locations in the CPU allowed modification of memory addresses and were particularly intended to help improve the efficiency of machine-code programs produced by compilers. Together with indirect addressing, the presence of index registers extended the memory addressing capabilities of computers in line with the requirements dictated by high-level languages. Experience with software was influencing the design of computers considerably. Applications, too, influenced their design. The requirement for very powerful computational facilities was satisfied by the widespread use of floating-point arithmetic units.

Details apart, the general structure of computers as specified by von Neumann and his colleagues was still the same — the three main parts, CPU, memory and input/output — and has remained so ever since. Moving on beyond 1960 the trend was still to improve the speed, size (and inevitably cost) and programmability of computers.

As with the previous generation, *third-generation* computers are most strongly characterised by a technological innovation, in this case the use of _integrated circuits_ (ICs). Instead of the former discrete components, the semi-conductor industry began producing monolithic ICs on which the equivalent of several transistors were fabricated. This newest advance took place in the early 1960s, with Fairchild and Texas Instruments well to the fore amongst the semiconductor manufacturers. It was, however, Sylvania which first produced the logic family which has remained popularly in use up to the present day: transistor-transistor logic (TTL). With higher packing density of components and improved switching speeds, ICs enabled computers (and other digital logic devices) to be much smaller and faster. It is alternatively suggested that third-generation machines are mainly characterised, from the programmer's point of view, by multiprogramming operating systems based on large capacity magnetic drum and disc stores. Certainly all of the major computer manufacturers set out to implement such operating systems, although it cannot be claimed that many had success until much later in the 1960s. One of the most successful third-generation machines was IBM's System/360, which was available in a variety of different configurations to meet the needs of the individual customer. These machines, in common with the majority being produced at the time, were very large, powerful _mainframe computers_.

About the middle of the 1960s a somewhat different type of machine began to appear on the market. This was the _minicomputer,_ characterised by short word lengths (of some 12 to 24 bits) and modest hardware and software facilities. These machines were built to satisfy a new, but soon growing demand for dedicated computers in industrial applications. Digital Equipment Corporation (DEC) was one of the first manufacturers, with its PDP series, to sell minicomputers.

Although more powerful computers continued to be designed, the trend towards smaller machines accelerated as whole new applications areas in industry and commerce revealed themselves. This trend was helped along considerably by the increasing performance/cost ratio of integrated circuit technology. In 1964 Texas Instruments introduced a standard TTL product line called semiconductor network (SN) series 54. Although this was intended primarily for the military market, Texas Instruments soon produced a lower-cost, lower-specification version called series 74. This logic family originally packaged up to about twelve transistor equivalents on one IC: this _level of integration_ is called small-scale integration (SSI). In 1969 medium-scale integration (MSI) was introduced, packing from twelve up to a hundred transistors onto an IC.

Large-scale integration (LSI) soon enabled upwards of a hundred transistors to be packaged together on one monolithic structure. With this level of integration manufacturers saw that they could produce an IC containing enough logic to implement a small CPU. In 1971 Intel brought the first _microprocessor_ into the marketplace, the 4-bit 4004. Soon 8-bit microprocessors, notably Intel's 8080 and the Motorola 6800, became very widely used products.

The increasing levels of integration and the lowering of IC component costs brought changes to computer memories as well as to CPUs. Semiconductor memories began to displace magnetic core as the standard memory product. These various innovations — LSI, microprocessors and semiconductor memories — have led us into, possibly, the *fourth generation* of computers.

Fig. 1.3 summarises the main characteristics of the computer generations. The dates are by no means universally agreed.

Generation	Characteristics
First (1945–1955)	vacuum tubes, delay line memory, paper tape/cards backing store, fixed-point arithmetic, machine-language programming
Second (1955–1965)	transistors, magnetic core memory, magnetic drum and disc backing store, floating-point arithmetic, high-level language programming
Third (1965–1971)	integrated circuits, magnetic core memory, magnetic disc and tape backing store, floating-point arithemtic, multiprogramming operating systems
Fourth (1971–)	large-scale integrated circuits, semiconductor memory, ...

Fig. 1.3 — Computer generations

Computer architecture

We have seen briefly how computer systems — hardware and software — have been developing up to the present day. Three broad classes of computers have been identified: mainframes, minicomputers and microcomputers (computers based round microprocessor CPUs). The distinction between these classes has traditionally been made by their application areas and also by what can loosely be termed their 'power'. This term reflects the number of program instructions a machine is capable of obeying per unit time; in addition, power is related to the hardware configuration — the number and the size of backing store and other peripheral units — which a machine is capable of supporting. Mainframes are traditionally powerful, whereas minicomputers are only moderately so. Microcomputers, up till now, have always been regarded as least powerful of all. We should perhaps no longer speak in such clear-cut terms about the three classes. The most recent 16-bit microprocessors are more powerful than many minicomputers, and strongly challenge some of the mainframes. It seems likely, however, that the size of a computer configuration will continue to determine its classification: a large, multi-access system will still be a mainframe.

The *architecture* of a computer is the structure and the inter-relationships of the various logical hardware units: the CPU, memory and peripheral units. This excludes the physical hardware construction of the surrounding machine like the power supply, the processor cabinet or any of the peripheral hardware. The physical construction and layout of the electronic circuits which make up the logical units may, however, be important in describing an architecture.

There are very many different computer architectures, but fundamentally they are all von Neumann machines. It is not the purpose of this book to discuss architectures, nor even any specific architecture, but rather to describe the fundamental building bricks — both logically and physically — of computers, and to show how they are used in the design and implementation of the units inside a generalised von Neumann machine. This is what is implied by *computer logic*. At the same time it must be stressed that although the treatment of logic elements and logic design is placed in the context of computers, the material is applicable in digital systems of all kinds.

To conclude this section let us outline a generalised von Neumann machine which reflects a typical modern computer architecture. This will be used as a basis for discussing the design of logical units within computers later in the book. Fig. 1.4 illustrates such a machine. Basically there are only three units —

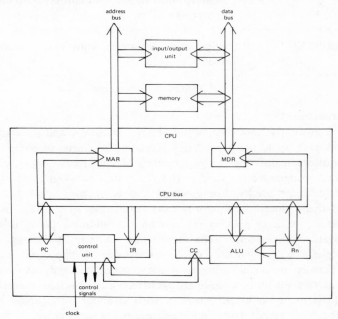

Fig. 1.4 — A generalised von Neumann machine.

CPU, memory and input/output (I/O) as previously described. In this diagram the units are shown connected by an _address bus_ and a _data bus,_ and the sub-units within the CPU (control unit and ALU) are shown with their associated _registers,_ connected by the _CPU bus._

The address bus is uni-directional: addressing information can be sent only from the CPU to one of the memory or input/output units. The data bus is bi-directional, meaning that data can pass in either direction, to or from the CPU. Inside the CPU information is transferred on the bi-directional CPU bus between registers — fast-access storage locations — as the CPU obeys program instructions. These registers require a little more explanation.

Associated with the address and data buses are the (memory) address register or MAR and the (memory) data register or MDR. In this generalised machine it is assumed that input/output units are addressed and data transferred to and from them in the way that applies to the memory unit. Such a system is called _memory-mapped I/O._

The length of or number of bits in these registers is important to note. Whereas the MDR length is the same as the basic _word length_ of the computer — that is the number of bits in parallel which the machine processes at once — the length of the MAR is related to the _addressing capability_ of the computer. It is quite common for a microprocessor to have a word length of 8 bits but an addressing capability of 64K, meaning that it is capable of reading or writing any of 64K different 8-bit storage locations. A MAR length of 16 bits gives an addressing capability of 64K. In general the relationship between MAR length and addressing capability is that

$$\text{addressing capability} = 2^{\text{MAR length}}$$

(independent of the basic computer word length).

The ALU has associated registers called Rn and CC. Rn stands for (any number of) general purpose registers, used for storing or accumulating information during the course of a program. These registers can also be used as index registers. CC is a condition codes register which records information about operations in the ALU, for example whether the result of an addition gave a positive, negative or zero result, or very importantly whether in an arithmetic operation the results register (one of the Rn) was not capable of holding the result — this is called overflow.

Lastly, the control unit has two registers PC and IR linked to it. These have a central role in the operation of the von Neumann machine, and to understand them it is necessary to describe how the computer runs programs. Every computer program, whether written in a high-level language or not, is eventually stored in the memory as a sequence of _machine-code instructions._ Each machine-code instruction implies a set of actions to be carried out by the computer to achieve the required goal. A typical instruction at this level is:

ADD Rn, X

meaning 'add the contents of memory location X to register Rn'. The binary form of this instruction (produced by the compiler or machine-code programmer) will occupy i consecutive memory locations. The number i depends on the type of instruction and also on the architecture of the computer. Each machine-code instruction in the sequence follows in successive memory locations.

The purpose of the PC or program counter is to hold the memory address of the current instruction as the program proceeds. Basically a computer fetches, decodes and obeys each machine-code instruction in a program. Each instruction, when fetched from the memory, has to be decoded by the control unit: the purpose of the IR or instruction register is to hold the current instruction while it is obeyed. As well as obeying the current instruction, the control unit has to alter the contents of the PC to point to the next instruction in sequence. For most instructions (such as ADD) the 'next' instruction is the one immediately following the current one. However, JUMP or BRANCH instructions are intended to cause the program to continue with a different part of its sequence (in conditional BRANCH cases, only if a specific condition, recorded in the CC register, is satisfied) and their only action in fact is to alter the contents of the PC to point to the 'next' location specified in the instruction.

The actions implied by a machine-code instruction are generally called *micro-operations*, small steps which the hardware can perform as a result of signals from the control unit. Most micro-operations correspond to the transfer of data from one register to another within the CPU. These micro-operations are called *register transfers* and are very important in the design and specification of computer architectures. A simple but effective notation for representing register transfers is shown in the example

$$PC \rightarrow MAR$$

meaning that the contents of the PC are copied into the MAR. This implies that the contents of the PC are not lost. Sometimes it is necessary to indicate indirect addressing, so for example

$$(MAR) \rightarrow MDR$$

means that the contents of the location pointed to by the MAR are copied into the MDR. In this case the MAR contents specify a memory location, and it is the contents of this memory location which are transferred into the MDR. The above operation is used every time a word of program or data is fetched from the memory. There is a corresponding register transfer for storing values into memory, namely

$$MDR \rightarrow (MAR)$$

where again the MAR specifies which location in memory is to be involved in the transfer.

To illustrate how a computer processes a single machine-code instruction let us consider the micro-operations required to implement the instruction ADD Rn, X ('add the contents of memory location X to register Rn'). It is assumed that the PC is already pointing to this instruction, and that in this case the instruction occupies two memory words, the first for the instruction code and the identity of the register, Rn, and the second containing the address X. Fig. 1.5 illustrates a micro-instruction implementation using the generalised von Neumann machine in Fig. 1.4. Note that the only micro-operation shown which is not a register transfer is (2), the decode step. The operation

$$MDR + Rn \rightarrow Rn$$

involves not only a register transfer, but also an addition within the ALU. Each micro-operation is initiated by signals from the control unit which, in the case

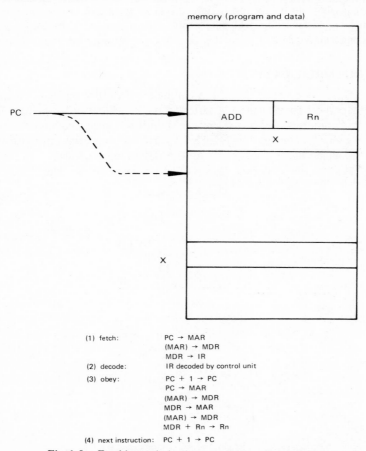

memory (program and data)

PC

| ADD | Rn |
| X | |

X

(1) fetch:	PC → MAR
	(MAR) → MDR
	MDR → IR
(2) decode:	IR decoded by control unit
(3) obey:	PC + 1 → PC
	PC → MAR
	(MAR) → MDR
	MDR → MAR
	(MAR) → MDR
	MDR + Rn → Rn
(4) next instruction:	PC + 1 → PC

Fig. 1.5 — Fetching and obeying a machine-code instruction.

of register transfers, open and close appropriate data paths, to connect the registers involved in the transfer. The fetch step fetches the first instruction word from memory. If, as in this example, an operand address is specified in the instruction it is fetched in the obey step. Steps (1) and (2) are identical for all instructions but (3) and (4) depend on the nature of the instruction. In the example ADD Rn, X there are altogether three memory fetches.

The control unit in turn is actioned by a *clock* or source of timing signals (usually external to the CPU) which determines the basic time for a micro-operation. Roughly speaking, each clock 'tick' causes the next micro-instruction to begin. The time increment between ticks (the *machine cycle* time) is chosen to allow the logic circuits to complete their operation: since micro-operations may take different lengths of time, the clock frequency or rate of ticking depends on the slowest operation. Obviously the faster the circuits operate, the faster the clock frequency can be and ultimately the more powerful the computer. Some machine-code instructions will take more, and others fewer, machine cycles in direct correspondence to the number of micro-operations required to implement them.

1.3 STRUCTURAL LAYERS

Computer logic is a subject which, in common with other topics in computing, can perhaps be best viewed as a series of layers which represent its essential structure. Computers are built from logical units which are themselves constructed from more fundamental building bricks. The building bricks themselves have a structure which the student of this subject should know about.

In Fig. 1.6 the structural layers of computer logic are illustrated. Three

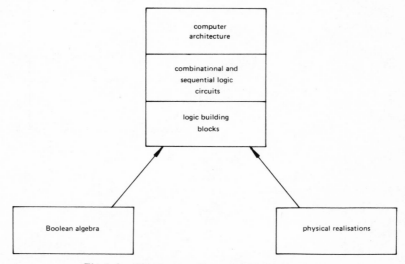

Fig. 1.6 — The structural layers of computer logic.

layers are identified along with the roots of computer logic in mathematical theory and the physics of circuit implementations.

The structure of this book reflects the layers in the diagram. Chapter 2 deals with logic building blocks and the background material of Boolean algebra and physical realisations, Chapters 3 and 4 with combinational and sequential logic, and Chapter 5, fairly briefly, with computer architecture. The last chapter takes the discussion a little way into the software part of the computer spectrum, to emphasise that hardware design and software requirements are closely inter-dependent.

Finally, the three layers can be seen to correspond roughly to what we may call the bit, word and processor levels of computer logic. In terms of available integrated circuit components the layers may also be associated with SSI, MSI and LSI respectively. These two sets of equivalences may on the one hand be regarded simply as alternative shorthand names for the layers, but on the other hand they should both be remembered because they reflect the presence of the abstract and physical streams which run through computer logic.

CHAPTER 2

Logic building blocks

2.1 LOGIC SYMBOLISM

Computer hardware, and digital systems of all kinds, are made of logic circuits. In practice there are many levels of complexity of logic circuits, although for the present it will be convenient to regard them as being all at the *word* level. Logic circuits in turn are built from more basic units at the *bit* level. This chapter is about the bit-level logic units which are the building blocks of digital logic circuits.

There are two distinct streams leading up to a full appreciation of logic building blocks, the abstract and the physical. These correspond to the *design* and *implementation* stages in which logic circuits are produced. The design process may be carried out in terms of abstract logic units without regard to the physical details of implementation. Of course the design must then be realised in hardware, using available components. It is always desirable to cut the cost and physical size of circuits to a minimum, while also ensuring that the speed of their operation is as high as possible. These *minimisation criteria* generally conflict with one another because of natural physical limitations, so a suitable balance has to be struck between the various criteria in the designer's mind. The choice must be made at the design stage so that the circuit structure may be manipulated appropriately. Because of the minimisation requirement, and also for the reason that circuit implementations depend on the availability of components, design and implementation must be closely linked in practice. Let us consider these two streams together.

A logic unit may conveniently be viewed as a *black box* as illustrated in Fig. 2.1. A black box is characterised by the fact that nothing is known, nor need be known, of its internal structure. At any time t, $I(t)$ represents the set if input values and $S(t)$ the *internal state* of the unit. A little time later, at t', the outputs become $O(t')$ and the internal state changes to $S(t')$. The outputs are derived from the inputs and internal state according to the transformation:

$$O(t') = f(I(t), S(t)) \qquad (2.1)$$

Similarly the new internal state is a function of the inputs and previous internal state:

$$S(t') = g(I(t), S(t)) \qquad (2.2)$$

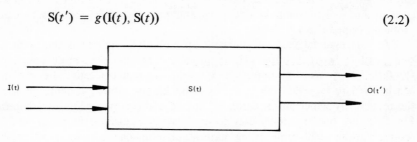

Fig. 2.1 – A black-box logic unit.

Each input and output has a finite number of possible values, and the unit a finite number of internal states. It is more generally known as a *finite state machine* (FSM). Although formal FSM theory is beyond the scope of this book, it is important to point out the close relationship between computer logic and the well-established work on FSMs (otherwise known as automata). The theory describes the structure and behaviour of FSMs in general, including the necessary and sufficient properties which enable them to be used as universal computing machines. Computer logic makes exclusive use of a class of FSMs in which individual inputs and outputs may have only two possible values.

The (simplified) FSM shown in Fig. 2.1 represents a general type of logic unit called a *sequential logic element*. Its characteristic is that its outputs depend on the present input values, and also on the past history of the inputs which are summarised by the present internal state. Although both equations (2.1) and (2.2) are important in describing the behaviour of the sequential element, only the inputs and outputs are actually observable. Note that because of the possibility that the element can be in one of a number of possible internal states at any one time, and because the outputs depend (partly) on the present state, the same input values applied at different times may produce different output results. An important special case of the sequential logic element is the one in which there is only a single internal state. This type is called a *combinational logic element.* The characteristic feature of a combinational element is that a given set of input values always produces the same output results. It can be described by a modified version of equation (2.1) in which no internal state is specified:

$$O(t') = f(I(t)) \qquad (2.3)$$

In this case there is no equivalent of equation (2.2) since the element has no alternative states to sequence through. Note that equation (2.3) implies a *delay* in producing new outputs whenever a new set of input values is applied to the element. Although for most of the purposes of designing logic circuits this delay can be ignored, it is a realistic factor which must be taken into account in some circumstances. This point will be discussed more fully in Section 4.2 ('Circuit problems in practice').

Another special case of the general sequential element is one with only two possible internal states. These are called *bistables* or, more colourfully, *flip-flops*. They are very important in computer logic, because the two states can be used to represent the 1s and 0s of the binary number system, the basis of the method by which information is handled and represented inside computers. Flip-flops are fundamentally easy to implement because of the many two-state representations achievable using natural phenomena: hole/no hole, current/ no current, high voltage/low voltage. They are most importantly used as *memory elements*.

There are many possible combinational logic elements, each distinguished by the function *f* which maps inputs to outputs as in Fig. 2.1, and by the number of inputs and outputs themselves. The most fundamental combinational elements are called AND, OR, NOT *gates*. The first two, AND and OR, are single-output elements but may be defined for any number of inputs, from two upwards. The third, NOT, is a single-output, single-input element. The NOT gate and two-input versions of AND and OR are defined in Fig. 2.2 alongside the black-box equivalent of each.

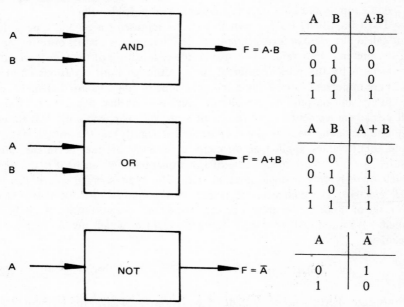

A	B	A·B
0	0	0
0	1	0
1	0	0
1	1	1

A	B	A + B
0	0	0
0	1	1
1	0	1
1	1	1

A	$\overline{A}$
0	1
1	0

Fig. 2.2 – The fundamental combinational logic gates.

The inputs and outputs of all logic functions can take only two possible values — the binary digits 0 and 1 — and thus it is particularly easy to describe these functions by tabulating all possible combinations of input values and writing down the corresponding output value in each case. The definitions of AND, OR, NOT in Fig. 2.2 are in this form which is generally known as the *truth table*. The term 'truth' derives from the origins of the theory of computer logic in which the two values of inputs and outputs are TRUE or FALSE. This will be elaborated in Section 2.2 following.

Fig. 2.2 shows the usual form of writing AND, OR, NOT operations: AND is represented by · ('dot'); OR by + ('plus'), both of these operators being placed between their two operands; and NOT by ⁻ ('bar') placed over the top of its single operand. It is important not to confuse the logical + as defined for OR with the usual arithmetical +. Operations AND, OR, NOT can also usefully be defined in plain words:

A·B has the value 1 only if A *and* B both have value 1; otherwise the result is 0.

A + B has the value 1 if either A *or* B has value 1; if both are 0 the result is 0.

$\bar{A}$ negates the value of A: if A is 0 then $\bar{A}$ is *not* 0 (that is, 1); if A is 1 then $\bar{A}$ is *not* 1 (but 0).

The OR operation + is more precisely called *inclusive-OR* to distinguish it from another operation called *exclusive-OR*. This is defined, and its operator symbol shown, in Fig. 2.3.

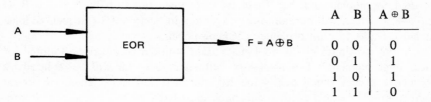

A	B	A ⊕ B
0	0	0
0	1	1
1	0	1
1	1	0

Fig. 2.3 – The exclusive-OR gate.

In plain words:

A ⊕ B has the value 1 if either A *or* B has value 1, but not both; if both are 0 or both are 1 the result is 0.

Although EOR is not fundamental (in the sense that it *can* be synthesised from AND, OR, NOT gates) it is nevertheless a commonly-used logic element and will be referred to again.

Two other very important functions related to AND, OR, NOT are NAND and NOR. Basically these are concentrations of AND — NOT and OR — NOT respectively, and are much used in practice. They are conveniently fabricated in integrated circuit form and are more readily available components than AND and OR. The definitions of NAND and NOR are shown in Fig. 2.4. Note that the 'bar' covers the entire function in each case, that is the NOT is applied to the output of the functions, not to their inputs.

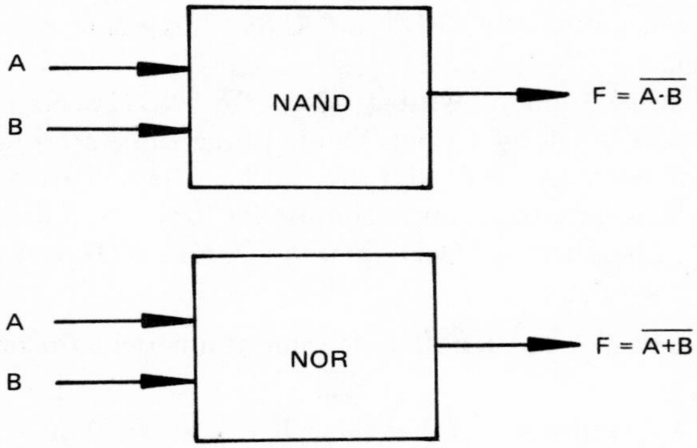

A	B	$\overline{A \cdot B}$
0	0	1
0	1	1
1	0	1
1	1	0

A
B
NAND
$F = \overline{A \cdot B}$

A	B	$\overline{A + B}$
0	0	1
0	1	0
1	0	0
1	1	0

A
B
NOR
$F = \overline{A + B}$

Fig. 2.4 – The 2-input NAND and NOR gates.

We have identified the various combinational logic gates – AND, OR, NOT, EOR, NAND, NOR – and sequential elements called flip-flops which are all logic building blocks. The fundamental operations AND, OR, NOT are used mainly in the abstract design of logic circuits, while NAND, NOR and NOT are widely available in small-scale integrated (SSI) form (EOR in MSI) and tend to be used for implementation. There are three main types of flip-flop – S-R, D and J-K – which will be more fully explained in Section 3.3 ('Sequential logic design'). These, too, are available as SSI components.

In the remainder of the book a standard symbolic form will be adopted for these logic elements. The black-box outlines used so far will be replaced by *distinctive shapes* which clearly identify the logic functions. There are several standards but the one used in this book is perhaps the most common. It conforms to the recommendations in 'IEEE Standard Graphic Symbols for Logic Diagrams' (IEEE Std. 91-1973, ANSI Y 32.14-1973). Drawing templates containing the recommended shapes are available commercially. Fig. 2.5 shows the distinctive shapes for all the combinational logic gates and one flip-flop, the S-R. All flip-flop symbols have the same basic rectangular shape but there are extra distinguishing features for other types which will be described in Section 3.3.

The basic logic building blocks have been introduced and their symbols illustrated. To be able to use them in designing and implementing logic circuits requires further development of their abstract and physical background. These two areas are explored further in the following sections of this chapter.

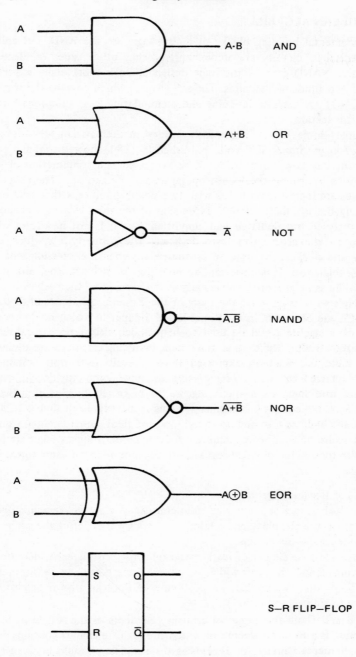

Fig. 2.5 – Distinctive logic symbol shapes used in this book.

2.2 BOOLEAN ALGEBRA

The fundamental building blocks of digital logic are the AND, OR and NOT gates. Although circuits are implemented using other types of elements — particularly NAND gates — the logic design phase is most easily expressed in terms of the fundamental gates. The set of rules which governs the way AND, OR and NOT are used in designing and manipulating logic functions forms the basis of this section.

George Boole was a nineteenth-century mathematician who devised an *algebra of propositions*. His work, published in 1854, showed that any proposition, no matter how complex, could be expressed as a combination of simpler propositions linked by the logical operators *and, or* and *not*. The propositions themselves are logical expressions with two possible values, either *true* or *false*. Boole's algebra is a mathematical expression of the rules by which propositions can be analysed, manipulated and simplified, but it is not in a form which is palatable to the prospective logic designer. The property by which the set {*and, or* and *not*} can be used to synthesise any logical expression whatever is called *completeness*. It is important to note that there exist other sets of operators with the same property, particularly {*nand*} and {nor} themselves.

Boole's work is however the basis of the modern techniques of digital logic design. Claude Shannon, in 1938, published a paper in which he demonstrated how Boole's algebra could be used to help design electromechanical relay circuits systematically. Up to that time such switching circuits were designed by ad hoc methods. Shannon expressed Boole's results in a more suitable form which is referred to as *switching algebra* or more commonly *Boolean algebra*. The basic usefulness of Boolean algebra as suggested by Shannon lies in its ability to enable complex switching systems to be expressed then optimised in terms of the basic *and, or* and *not* operators, and finally implemented using their physical realisations. The techniques used for modern logic design are just the same, even though the physical realisations may not be in the same form.

The rules of Boolean algebra

Boolean algebra, like any other algebra, consists of a *set of elements, operators* which act on the elements, and a number of *rules* which define the properties of both elements and operations.

There are two classes of rules: basic rules (or *postulates*) which are stated without proof, and *theorems* which are proved using postulates and/or previously proved theorems. In Fig. 2.6 the postulates of Boolean algebra are listed, in a form due to Huntington (1904).

A, B and C are the names of arbitrary elements of the set. In logic design our interest is a Boolean algebra in which the set of elements contains only the identity elements namely {0, 1}. This is the simplest possible Boolean algebra. Any *variable* such as A, B or C can therefore take only the values 0 or 1.

P1 There are two operations · and + on pairs of elements in the set which produce a result also belonging to the set of elements: this is the *closure rule*.

P2 The operations · and + are *commutative:*

$$A·B = B·A$$
$$A+B = B+A$$

P3 The operations · and + are *distributive:*

$$A·(B+C) = (A·B)+(A·C)$$
$$A+(B·C) = (A+B)·(A+C)$$

P4 Two elements 1 and 0 called *identity* elements exist such that:

$$1·A = A$$
$$0+A = A$$

P5 For each element A in the set there is an *inverse* $\overline{A}$ such that:

$$A·\overline{A} = 0$$
$$A+\overline{A} = 1$$

Fig. 2.6 – The postulates of Boolean algebra.

The postulates state the basic properties by which we shall be able to manipulate logic expressions. A number of theorems which will also be found useful for this purpose are listed in Fig. 2.7. Two theorems in particular are very important: T9 and T10, the so-called *de Morgan's laws* named after their originator. These tell us how the *inverse* of arbitrary logic expressions can be rewritten in terms of the inverted variables in the expression. Although T9 and T10 show a two-variable case the theorems apply also to expressions containing any number of variables greater than two. For example

$$\overline{A·B·C} \qquad = \overline{A} + \overline{B} + \overline{C}$$

and $$\overline{A+B+C+D} = \overline{A}·\overline{B}·\overline{C}·\overline{D}$$

demonstrate the 3-variable form of T9 and the 4-variable form of T10. De Morgan's laws also provide a means by which · operations can be changed to +, or

vice versa, by forming the inverse of an expression. This can be useful for manipulating logic expressions into a form suitable for implementation: in particular
it is often required to implement circuits entirely using NAND gates, so basically
any + operations in the original expression of the circuit to be changed to ·
operations.

T1	$A \cdot 0$	$= 0$
T2	$A+1$	$= 1$
T3	$A \cdot A$	$= A$
T4	$A+A$	$= A$
T5	$A+(A \cdot B)$	$= A$
T6	$A+(\bar{A} \cdot B)$	$= A+B$
T7	$A \cdot B \cdot C$	$= (A \cdot B) \cdot C = A \cdot (B \cdot C)$
T8	$A+B+C$	$= (A+B)+C = A+(B+C)$
T9	$\overline{A \cdot B}$	$= \bar{A}+\bar{B}$
T10	$\overline{A+B}$	$= \bar{A} \cdot \bar{B}$
T11	$\overline{(\bar{A})}$	$= A$

Fig. 2.7 – The theorems of Boolean algebra.

Theorems T7 and T8 are called the *associative laws*. Together with the
commutative and distributive laws (postulates P2 and P3) they describe properties of · and + which are very similar to the properties of the arithmetic operators in the everyday algebra of real numbers. For this reason we find · and +
convenient operators to work with in abstract logic design (rather than, say,
the NAND and NOR operators). The + operator, pronounced 'plus', is also
called the *sum* operation, while · ('dot') has the alternative title *product*. Resemblances to the arithmetic operators are rather superficial and the parallel must
not be taken too seriously. An abbreviation arising out of the resemblances,
however, is widely used: the · operator may be omitted altogether in logic
expressions. For example theorem T7 may be written as

$$ABC = (AB)C = A(BC)$$

The + operator always appears explicitly. Note the use of brackets in theorems
T5 and T6. Bracketed expressions show explicitly how the expression is to be
evaluated — the expressions in brackets first. However, brackets may be omitted
from expressions, in which case the *implicit strengths* of the operators determine
the order of evaluation. The important rule is that · is stronger than +. Thus
theorem T5 can equally well be written:

$$A + A \cdot B = A$$

or indeed $A + AB = A$

Proving the theorems

Two methods may be used for proving the theorems of Boolean algebra. The first, *perfect induction,* involves evaluating the expressions we are trying to prove equal for all possible values of the variables, and comparing the results in each case. This method is particularly easy for Boolean expressions because each variable has only two possible values. Fig. 2.8 gives an example of this type of proof for theorem T5.

A	B	A·B	LHS: A + A·B	RHS: A
0	0	0	0	0
0	1	0	0	0
1	0	0	1	1
1	1	1	1	1

Fig. 2.8 – Proof of theorem T5 by perfect induction.

The second method is by *algebraic manipulation* whereby the postulates (and previously proved theorems) are employed in re-writing the expressions on one or both sides until they are identical. An example of the algebraic proof for theorem T6 is given in Fig. 2.9. Each re-writing of an expression should be justified by quoting the identity of the postulate or theorem used.

$$
\begin{aligned}
\text{LHS:} &= A+(\overline{A}B) \\
&= (A+\overline{A}) \cdot (A+B) & \text{by P3} \\
&= 1 \cdot (A+B) & \text{by P5} \\
&= (A+B) & \text{by P4} \\
&= \text{RHS}
\end{aligned}
$$

Fig. 2.9 – Proof of theorem T6 by algebraic manipulation

Duality

The postulates (apart from the closure rule P1) consist of pairs of expressions. These are called *dual forms* and reflect the symmetry of the · and + operators. Every Boolean expression has a dual which may be derived by replacing each occurrence of · by +, each + by · , 0 by 1 and 1 by 0. Inversion is not affected. Notice that in Fig. 2.7 the dual forms are paired consecutively: T1 and T2 are duals, also T3 and T4, T5 and T6, followed by the associative laws (T7 and T8), and de Morgan's laws (T9 and T10). For obvious reasons T11 appears on its own.

A note on Boolean expressions

The Boolean expression

$$AB + CD$$

is said to be in a *sum-of-products* form, where AB and CD are product terms linked by the sum operator. The expression has a dual form

$$(A+B)(C+D)$$

known as a *product-of-sums* form. In logic design it is usual to adhere to one or other of the two forms throughout the process of expressing and manipulating Boolean expressions. In this book we shall use the sum-of-product form for writing down logic functions.

A general Boolean function of n variables may be expressed in a sum-of-products form as follows:

$$f(A_1, A_2, \ldots, A_{n-1}, A_n) = \bar{A}_1 \bar{A}_2 \ldots \bar{A}_{n-1} \bar{A}_n f(0,0, \ldots, 0,0)$$
$$+ \ \bar{A}_1 \bar{A}_2 \ldots \bar{A}_{n-1} A_n f(0, 0, \ldots, 0, 1) \ +$$
$$+ \ \bar{A}_1 \bar{A}_2 \ldots A_{n-1} \bar{A}_n f(0, 0, \ldots, 1, 0) \ + \ \ldots \ldots \ldots$$
$$+ \ A_1 A_2 \ldots A_{n-1} A_n f(1, 1, \ldots, 1, 1)$$

(giving $2\uparrow(2\uparrow n)$ n-variable Boolean functions).

For $n = 2$ we have:

$$f(A, B) = \bar{A}\bar{B}f(0,0) + \bar{A}Bf(0,1) + A\bar{B}f(1,0) + ABf(1,1)$$

The values of $f(0, 0), f(0, 1), f(1, 0)$ and $f(1, 1)$ can be either 0 or 1, giving a total of 16 different Boolean functions of two variables as shown in Fig. 2.10. The familiar functions are identified by their names.

The specific case of the exclusive-OR function EOR can be written as:

$$f6 = A \oplus B = \bar{A}\bar{B}{\cdot}0 + \bar{A}B{\cdot}1 + A\bar{B}{\cdot}1 + AB{\cdot}0$$

but since by the rules of Boolean algebra

$$A{\cdot}0 = 0 \quad (T1) \quad \text{and} \quad A{\cdot}1 = A \quad (P4 \text{ and } P2)$$

the expression reduces to $A \oplus B = \bar{A}B + A\bar{B}$.

This sum-of-products form can be derived more simply from the truth table (see Fig. 2.10 or Fig. 2.3) by identifying the combination of input variables which produces each 1 output. The combinations are expressed as product terms and summed together. The final expression for EOR may be interpreted in words: the output is 1 if (*not* A) *and* B is 1 *or* A *and* (*not* B) is 1.

B	0	1	0	1	
A	0	0	1	1	
$f0$	0	0	0	0	
$f1$	0	0	0	1	AND
$f2$	0	0	1	0	
$f3$	0	0	1	1	
$f4$	0	1	0	0	
$f5$	0	1	0	1	
$f6$	0	1	1	0	EOR
$f7$	0	1	1	1	OR
$f8$	1	0	0	0	NOR
$f9$	1	0	0	1	
$f10$	1	0	1	0	
$f11$	1	0	1	1	
$f12$	1	1	0	0	
$f13$	1	1	0	1	
$f14$	1	1	1	0	NAND
$f15$	1	1	1	1	
	$f(0,0)$	$f(0,1)$	$f(1,0)$	$f(1,1)$	

Fig. 2.10 – The Boolean functions of two variables.

2.3 LOGIC FAMILIES

Boolean algebra provides the tools by which abstract logic functions can be expressed and manipulated. Paper designs are not an end in themselves and must be closely linked to the implementation stage in which logic circuits take their final physical form. The implementer is faced with choosing from the available physical building blocks, of which there are increasingly many. This section, and the one following, are concerned with the physical realisations of abstract logic elements and aim to provide a background of knowledge about their characteristics which will aid the implementer in his choice.

Several alternative ranges of electronic components with which to build computers, and other digital devices, have been developed since the first-generation machines. All computers since then have been binary (two-valued) in principle, working in terms of high or low voltages, current flow and lack of current, and (for the memory) the two opposite directions of magnetisation. At the very lowest level of operation is the basic *electronic switching element* which allows the implementation of two-valued quantities. Since 1948 this switch has been, in one form or another, the *transistor*. Corresponding to the abstract design level, the fundamental building block for implementation is the *gate,* a circuit based round transistors which provides one of the logical operations previously described (AND, NOR, etc). It is the form of the basic gate which distinguishes each of the range of components or *logic families.*

The early logic families, including *diode transistor logic* (DTL) and *resistor transistor logic* (RTL), provided gates which were individually packaged. These families are now of mainly historical interest. Development of the *transistor-transistor logic* (TTL) gate in the 1960s enabled a number of gates of the same type to be packaged together in one integrated circuit (IC) for a cost little greater than that of the previous, discrete gate. The gates were fabricated on a *monolithic* structure, that is a single chip of semiconductor material (hence the popular name *chip* for a packaged IC). Our interest in logic families starts chronologically with TTL.

At first, only a small number of gates could be fitted into one package. Fabrication techniques have been developing continually since then and the complexity of packaged circuits has correspondingly increased. Three levels of integration are currently identified: SSI, MSI, and LSI (small, medium and large scale integration). These correspond roughly to complexities of 10, 100 and 1000 gates in one IC. More recently gate complexities of the order of 10 000 have been achieved; the name reserved for this new level is VLSI (V meaning very). In contrast with SSI chips which mostly contain several separate gates of the same type, MSI and the higher-level ICs consist of single, complex logic circuits. Examples of these are: MSI–registers, adder units; LSI–multiplier units, memory chips, microprocessors; VLSI--large scale memory chips, complex microprocessors. All these ICs are intended for use as building blocks of digital systems. Since TTL came into being the variety of available ICs has increased enormously, particularly in the medium complexity range. The trend in building systems is to use the highest-level ICs available for each part of the design. Often, these parts have to be interlinked by logic at the gate or SSI level. With larger-scale integration tends to come generality in the function of the ICs: if one of the LSI units does not provide quite the function which the system designer requires, nor is it capable of being modified by extra logic, then the system must be constructed from the next-lower level, MSI, building blocks.

Physical realisation of gates

At the heart of every logic family is the fundamental electronic switching element, the transistor. Some of the families are based on the *bipolar transistor* invented in 1948 by Bardeen and Brattain, and others on the *unipolar transistor* first described by William Shockley in 1952. The difference between the two types lies in their detailed structure and semiconductor action – giving the associated logic families characteristic properties. Broadly speaking the transistor structure and action are the same for both transistor types, and may adequately be described as follows by using the bipolar transistor as an example.

Transistors are made from semiconductor material such as germanium or silicon which have an intrinsic electrical conductivity between that of conductors and insulators. Their conducting properties may be altered by doping the material with either n-type or p-type impurities. The result of n-type doping is that a suitable voltage applied to the material causes *electrons* (negative charge carriers) to flow whereas p-type doping results in a flow of positive charge carriers called *holes*. The bipolar transistor – or junction transistor as it is sometimes called – is formed by diffusing n-type and p-type impurities into the same piece of semiconductor material so that two *p-n junctions* are formed, as illustrated in Fig. 2.11. This shows an n-p-n transistor structure together with its circuit symbol.

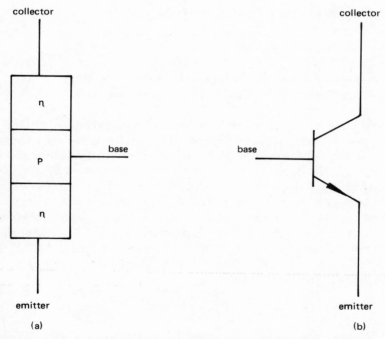

Fig. 2.11 – The bipolar transistor: (a) structure, (b) symbol.

Alternatively a p-n-p transistor, with similar properties, may be made from one n-type and two p-type junctions. The arrow on the symbol indicates the direction of conventional current flow: electrons flow in the opposite direction, originating from the emitter. They travel to the base from which most make their way to the collector. Holes produced at the base terminal are attracted to the emitter, although some combine with electrons inside the base region. The name bipolar reflects the fact that both types of charge carriers are active.

To explain the transistor action consider the *common emitter* configuration of Fig. 2.12. The circuit INPUT is the base, its OUTPUT the collector. This

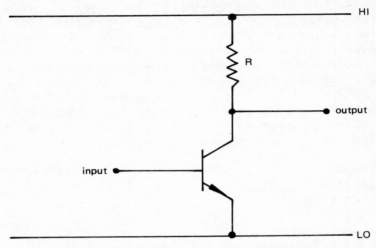

Fig. 2.12 – Common-emitter transistor configuration.

arrangement, in which the emitter is common to both INPUT and OUTPUT, is used in most switching applications. The collector is connected via a resistor R of suitable value to the positive voltage supply (HI) while the emitter is at zero voltage (LO). The operation of the transistor is summarised by the output characteristic graph shown in Fig. 2.13. This is a plot of the collector current I_c as a function of collector-emitter voltage V_{ce} and base current I_b. When there is zero base current – INPUT at ground potential – I_c is very nearly zero (a small *leakage* current flows between collector and emitter). Under these conditions the transistor is said to be OFF. By applying an increasing potential at the INPUT, at constant V_{ce}, the value of I_c increases in proportion to the increasing base current according to the relationship

$$I_c = hI_b$$

where typically h is 10 or more. The transistor *amplifies* the input current. When the INPUT voltage is increased to a suitably high value the collector-emitter potential drops to nearly zero and a large collector current flows. The transistor is then said to be *saturated* or in the ON state. Thus the transistor appears as a switch controlled by the base current (or voltage). In the OFF state the collector current is effectively zero and the OUTPUT potential is at HI; when ON the collector current is large and the OUTPUT is almost at LO.

Note that the circuit in Fig. 2.12 has an *inverting* action: INPUT at LO gives OUTPUT at HI, but INPUT at HI produces OUTPUT at LO. This circuit

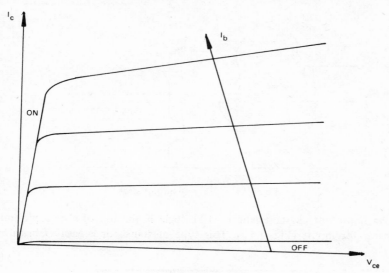

Fig. 2.13 – Transistor output characteristic.

basically implements a NOT gate. Its amplification properties (sometimes called *gain*) enable the output of the gate to control the inputs of several other, similar gates, an important requirement in logic circuits. The number of inputs which may be so connected to one output is called the *fanout* capability of the gate.

Although Fig. 2.13 demonstrates that the bipolar transistor, working in saturated mode, may be switched between the OFF and ON states, it gives no indication of the *switching time*. Of all the properties of transistor circuits this is perhaps the most important, since it determines the *speed* of the logic family. The time taken by a logic gate to switch output states when the input is changed from OFF to ON (or vice versa) is called the *propagation delay* of the gate. Many of the developments in IC technology have resulted from a desire to reduce gate delay. As a simple example, the switching speed of the circuit in Fig. 2.12 can be improved by using two transistors as illustrated in Fig. 2.14. This shows the output stage common to TTL gates, called an *active pull-up* or *totem-pole* output. The gate logic is specific to the type of gate but in each case ensures that one of two conditions applies at the output stage: either T1 is OFF and T2 is ON, in which case the OUTPUT is (almost) at HI (+5 volts for TTL logic), or T1 is ON and T2 is OFF giving a LO (0 volts) at the OUTPUT. The typical gate delay for this TTL configuration is 10 ns (ns = nanosecond = 10^{-9} second). D is a diode which helps keep the switching time low and also improves the electrical *noise immunity* of the device.

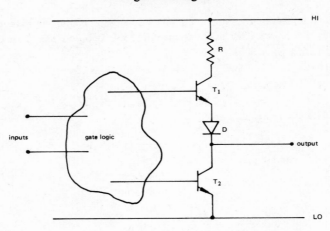

Fig. 2.14 — TTL gate output stage.

An important characteristic of TTL logic is the use of a multiple-emitter transistor for inputs (Fig. 2.15). This type of transistor is easily fabricated in

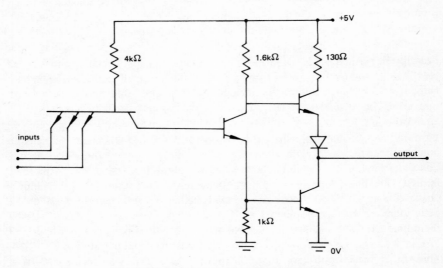

Fig. 2.15 — A 3-input TTL NAND gate.

monolithic form and gives a compact input stage. Fig. 2.15 shows the original series 54/74 TTL gate. It performs the NAND operation — a three-input version is shown but there are also 2-, 4- and 8-input versions. Because of its wide availability and because the NAND is itself a logically complete set, as explained in Section 2.2, many logic designers work entirely in terms of NAND elements.

We shall do the same in this book (apart from making convenient use of NOT gates).

Various forms of TTL logic have now been developed, all belonging basically to the same family. These are the standard (N) type as described above, and the low-power (L), high-speed (H), Schottky-clamped (S) and low-power-Schottky (LS) versions, each with its own special advantages (and complementary disadvantages). The Schottky varieties are examples of *non-saturated* mode devices. Diodes are added to the transistors in such a way as not to become saturated when ON and this improves the device switching speed.

Other logic families which use the bipolar transistor are available. *Emitter-coupled logic* (ECL), which also operates in a non-saturated mode, is currently the fastest logic family, but is particularly susceptible to noise and has a relatively high *power consumption*. Unfortunately high speed and low power consumption, the two most desirable logic family properties, are in general achieved each at the expense of the other. Their product, the so-called *speed-power product*, is often used to compare and classify logic families as we shall see later in this section. Both TTL and ECL are used in making SSI and MSI building blocks.

Integrated injection logic (I^2L) is a bipolar family, principally used in LSI components because of its small space requirements: thus chips with a high circuit density can be manufactured in this technology. The name derives from the use of a p-n-p transistor to inject current into the base of several active n-p-n switching elements. The injection device is merged into the structure of the n-p-n transistors. I^2L combines its low packing density with moderately high speed operation and moderately low power consumption.

The other logic families of interest are based on the unipolar or *field-effect transistor* (FET), reported in 1952 but not available in IC form until the mid 1960s. With one notable exception they are used in LSI production due to the much greater smallness of the FET in comparison with bipolar transistors. There is a variety of types of FET: broadly speaking they fall into two classes, called *depletion-mode* and *enhancement-mode*. The depletion-mode devices have current flowing in them in the absence of an input voltage, whereas conduction in enhancement-mode FETs is absent under these conditions and requires an applied input voltage to start current flow. Enhancement-mode FETs are used in practice in digital ICs: the structure and circuit symbol for this type of device are illustrated in Fig. 2.16.

Note that, just as with bipolar transistors, there are two kinds of construction: a piece of p-type material (the *substrate*) into which two n-type islands are diffused, or an n-type substrate with p-type islands. The illustrated type is the first of these, called a *n-channel* FET (the other is a *p-channel* device). The alternative name unipolar derives from the fact that only one kind of charge carrier is active in these devices, electrons in the n-channel and holes in the p-channel FET. Although p-FETs are easier to fabricate, they are slower than n-FETs because holes have inherently lower mobility than electrons. The most common fabrication method for both types is called the *metal-oxide-semiconductor* (MOS) technique. The gate (Fig. 2.16) is a metal electrode separated from the semiconductor material by a thin layer of silicon dioxide. The substrate is generally connected to the source. In operation as a switching device the MOSFET can be regarded as similar to the bipolar transistor (see Fig. 2.12), with gate, source and drain corresponding to base, emitter and collector respectively. The n-channel MOSFET of Fig. 2.16 is normally called an NMOS transistor (and its relative the p-channel, PMOS).

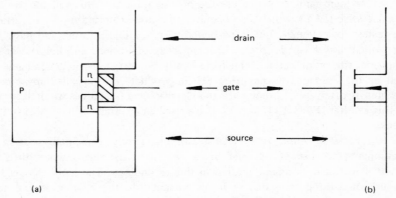

(a)   (b)

Fig. 2.16 – Enhancement-mode FET: (a) structure, (b) symbol.

Both NMOS and PMOS transistors are commonly used in LSI. Their advantages, in addition to being very small, are that they are simple and economical to fabricate and consume little power. They are, on the other hand, slow in operation compared to the bipolar families. In applications where speed is not of prime importance, NMOS in particular is a popular choice of technology.

A logic family closely related to NMOS and PMOS has been developed with low power consumption as the chief aim. Using both n- and p-channel FETs it is called *complementary MOS* (CMOS) and is the exception amongst unipolar logic families: it is used primarily for SSI and MSI components. The reasons for this are that CMOS devices occupy much larger areas than NMOS or PMOS and are more complex and costly to manufacture, so they cannot compete in LSI — but they can challenge TTL in the SSI/MSI area because of their greater packing density and much lower power consumption. Other advantages of CMOS include a wide range of operating voltages and a very large fanout capability. Perhaps the most important use of CMOS technology is in low power applications such as digital watches, which must run for long periods from battery sources.

The last logic family of interest is a unipolar family called *silicon-on-sapphire* (SOS), a variation of CMOS on which sapphire is used as an insulator to improve the switching speed of the logic. Although the fabrication process is very expensive SOS has significant benefits in both high speed and low power consumption. It is being increasingly used in LSI components despite its high cost.

A comparison of logic families

As previously mentioned the two most important aims for any logic family are high speed (that is low gate propagation delay) and low power consumption. To a great extent each of these tends to be achieved at the expense of the other. In comparing logic families it is useful to plot their speed-power products: this is done in Fig. 2.17.

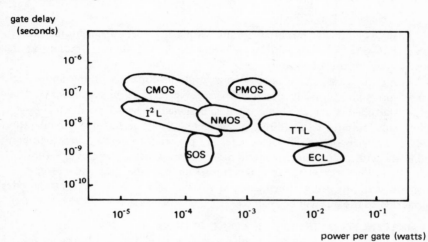

Fig. 2.17 – Comparison of logic families.

Essentially the product of gate delay and power per gate gives the amount of energy required for a switching operation. The lower the figure is, the better. On this basis, SOS would appear to be best all round, CMOS and I^2L are best for power consumption and ECL the fastest. It is, however, worth keeping in mind the other factors which contribute to the merit of a logic family, including packing density, fanout capability and the cost of fabrication.

The TTL family

Let us return briefly to the TTL family; despite the fact that other families have been developed in competition, TTL remains the most popular technology for SSI and MSI components. Any given logic family has a fairly small range within which its speed-power product can be varied, but either property can be improved at the expense of the other rather more easily. This has been exploited in TTL to produce five different types as mentioned earlier. Their characteristics are listed in Fig. 2.18.

Type	Power Per Gate (mw)	Gate Delays (ns)
Standard (N)	10	10
Low power (L)	1	33
High speed (H)	20	6
Schottky (S)	20	3
Low-power Schottky (LS)	2	10

Fig. 2.18 — The five types of TTL logic.

Their speed-power products are all within an order of magnitude of each other. Low-power Schottky is rapidly displacing the standard variety as the most popular of the TTL family.

Lastly there are three output configurations for each of the above logic types: *totem-pole* (already described — see Fig. 2.15), *open-collector* and *tri-state*. The open-collector arrangement is designed to overcome the inability of totem-pole outputs to be directly connected together in an AND operation (called the *wired-AND*). Tri-state gates are provided to permit circuits to communicate with each other over common buses. A third output state (not LO, not HI, but a *high impedance* condition) disables the gate and prevents undesirable currents from flowing: at any time only one of the tri-state gates connected to the bus will be enabled.

2.4 INTEGRATED CIRCUIT BUILDING BLOCKS

The growth of semiconductor industries has been very rapid since integrated circuits were first produced in the early 1960s. There has been an enormous demand for ICs particularly for the computer and computer-related markets. From the first SSI chips containing the basic gates the range of available ICs has exploded, so that now there are several logic families, between them offering components at all levels of integration to meet a variety of needs. The very variety of ICs is somewhat overwhelming, but they may be classified into two broad areas: *digital* and *linear*. Linear ICs include operational amplifiers, voltage regulators, analogue-to-digital (A/D) and digital-to-analogue (D/A) converters, phase-locked loops and consumer devices for radio and television. The operational amplifier is the predominant linear device. Although these ICs are used in computer systems, particularly in A/D and D/A interfaces, they are not our concern here.

Digital devices, by far the larger of the two classes, may be further subdivided into the following categories:

        logic
        memory
        interface

Not all the literature available from semiconductor manufacturers and their appointed distributors uses these classifications — for example 'logic' is sometimes called 'digital', implying that memory and interfacing components do not belong to this class. What do the three categories listed above include? They are chosen to correspond roughly with the three main areas of computer hardware — CPU, memory and I/O.

Integrated circuit packages

From the logic implementer's point of view the form in which ICs are packaged is very important. Part of the task in implementing a circuit design is to ensure that the circuit diagram can be easily translated into physical connections. Circuits are built on boards specially made for the purpose, with an array of holes drilled through them so that the components can be mounted and soldered directly, or instead sockets into which the components will be inserted later. The designer chooses a board of sufficient dimensions for the circuit and makes a plan showing the locations of various ICs, resistors, capacitors and so on. When the necessary components and sockets have been soldered into place all that remains to be done is to connect their inputs and outputs according to the circuit diagram. Connections are made either by wire and solder or by the method of wire-wrapping, for which special pins have to be inserted in the board (these may be part of the IC sockets already mentioned).

Whichever inter-connection technique is used it cannot be done without a *point-to-point wiring diagram.* Usually it is possible to incorporate this on the designer's circuit diagram, although for more complex circuits it may have to be made separately.

IC packages come in three forms:

TO
flat
dual-in-line (DIL)

The TO are cylindrical metal packages which have been developed from transistor cases. They come in a variety of sizes. Flat packages are useful in applications where the product must be as two-dimensional as possible, such as in wrist-watch manufacture. Most used in computer applications are DIL packages, which have the basic outline shown in Fig. 2.19.

The body of the package is either ceramic (hermetically sealed for demanding environments) or more usually the cheaper plastic form (suitable for most applications). The metal leads or *pins* are connected to the chip itself which is enclosed inside the package body. Each pin has a function which is specified on the *data sheet* provided by the manufacturer. Two of the pins (at least) are reserved for connection to the power supply: these are usually indicated by the names V_{cc} and GND. The pin configuration shown in the data sheet conventionally uses the top view of the package (as in Fig. 2.19(a)).

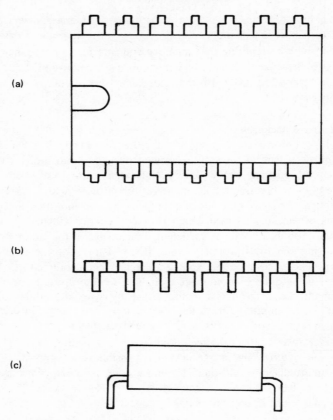

(a)

(b)

(c)

Fig. 2.19 – A DIL integrated circuit package: (a) top, (b) side, (c) end view.

Various sizes of DIL package are produced, the number of pins depending on the type of IC – although note that the pin spacing is always the same: 0.1 inches. SSI packages generally have 14 or 16 pins, but MSI and LSI types tend to be larger, with 20-, 24- and 40-pin packages. The newest microprocessors may have larger packages still, 48 pins or more. Packaging is very costly, so manufacturers tend to minimise the pin count for new designs of IC (even though more standardisation of pin positions, another desirable aim, can be achieved with larger packages, leaving some pins unused). The 40-pin micro-processors, for example, would be larger still had this minimisation not been applied.

Fig. 2.20 shows the pin configuration for a 2-input NAND gate (SSI) package. This package has 14 pins, of which two are used for the power supply lines. There are four identical gates in this package.

Note that the notch at the left hand end of the package in the diagram is

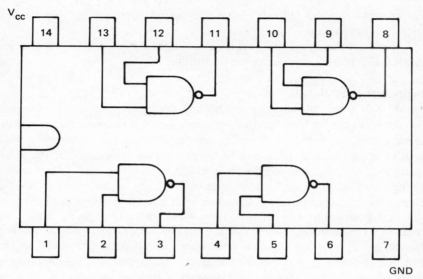

Fig. 2.20 – Pin configuration for 2-input NAND gate (SSI) package.

present on the package body itself: the pin numbers are not. Pin 1 is always
in the position shown, relative to the notch (see also Fig. 2.19(a)). For SSI,
and some MSI, ICs the function of each pin is pictorially described on the data
sheet. In the case of larger-scale integration chips the pins are identified by
name (simply a label beside each, with an annotated explanation).

For 3- and 4-input gates the package size stays the same but the number of
gates in each is reduced. NAND gates are available as follows:

$$4 \times 2\text{-input}$$
$$3 \times 3\text{-input}$$
$$2 \times 4\text{-input.}$$

The jargon in the literature calls these *quad, triple* and *dual* configurations
respectively. The package of 6 NOT gates is referred to as *hex*, while there is an
MSI package containing 8 flip-flops called *octal.*

All IC packages are identified by a unique *designator* or reference number.
There is a fairly good standardisation between manufacturers, so that the same
type of package is likely to bear the same reference number, with the manufacturer's symbol or name beside it. The IC in Fig. 2.20 is a 7400, in fact the
first in the 54/74 TTL series produced by Texas Instruments. As new packages
are marketed so they are given a higher number, but still starting with 74 to
identify the logic family. When ordering or specifying ICs it is important to
add to the designator a suffix specifying which type of packaging is required.

For example N and J indicate plastic and ceramic DILs respectively, while W refers to a flat package.

The Appendix (page 193) shows a selection of typical literature and data sheets available from a manufacturer or distributor.

Digital integrated circuits

Interface ICs are intended for connecting peripheral devices to computers. Some devices are for local interfacing, others for remote connection over transmission lines. In general the devices are for *line driving* or *receiving* where high power is required (compared to the power capabilities of logic ICs, inadequate for that purpose), and the environment can be very noisy. They tend to be mixtures of linear and digital circuits. Recently general-purpose interface chips have been produced in conjunction with microprocessor chips (and memory units) to provide a kit of parts from which microcomputer systems can be built in a standard way. Both serial and parallel devices are available, which go by the names of ACIA (asynchronous communications interface adapter) and PIA (parallel interface adapter) respectively. These are both memory-mapped devices, meaning that they are addressed as if they are memory locations but actually allow the transmission of data to and from peripheral units connected to them. In these two cases the drive capability of the chips is limited so for heavy-load or remote interfacing other more suitable chips have to be selected.

An increasingly important area in computing is that of *distributed* systems, networks of computers and other devices connected together by data transmission lines. Interface devices are likely to increase in variety and importance as these systems become more widespread. They will normally be designed to adhere to the interfacing standards being set up by the telecommunications and computing communities so that difficulties in connecting equipment from different manufacturers will be minimised. As the X-series recommendations for data communications become fixed then adopted, so complex interfacing chips to implement them will be manufactured.

Memory ICs consist of arrays of semiconductor memory cells, each cell capable of storing one bit of information. There is an enormous variety of memory chips of different capacities, speeds and configurations. The constant decline in the price of computer hardware has been very largely due to the falling cost per bit of memory. Semiconductor manufacturers have always been keen to pack more bits onto a single chip, and there is fierce competition to be the first to produce commercial quantities of yet more densely packed ICs.

The two main classes of IC technology are used in fabricating memory technology. Bipolar memories are faster but cannot be so densely packed: MOS (predominantly NMOS) memories yield large packing densities but do not operate so fast. For low-power applications CMOS memories are available.

Computer memories consist of groups of bits called *words,* each word being identified by a unique numerical *address*. The dimensions of the memory depend on the number of bits in the address and data paths of the CPU (as described in Section 1.2). <u>A memory space with addressing capability m, and n-bit word length (an $m \times n$ memory), is illustrated in Fig. 2.21.</u>

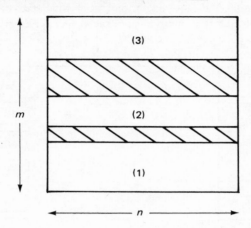

Fig. 2.21 – A fragmented $m \times n$ memory space.

The actual memory size may be any value up to m. It is increasingly common for the memory to be *fragmented* within the address space, as the diagram indicates. Areas (1), (2) and (3) are the occupied regions; the shaded areas are empty. It is now quite common to find different types of memory sharing the same address space, as will be explained below.

Common word lengths n used in computers range from 8 to 32 bits (and sometimes more). The majority of computers use either 8-, 16-, 24- or 32-bit words. This fact is reflected in the *organisation* of available memory chips (their dimensions in terms of m and n), the most common n values being 1, 4 and 8 bits: chips are often quoted only by their total storage capacity, for example '4K' (=4096) bits, but it is important also to know how the cells are organised internally. A 4K chip might be a 4K $\times$ 1, or a 1K $\times$ 4, or even 512 $\times$ 8. Even though these three configurations represent the same total number of bits they are quite distinct from the designer's point of view. Each requires an IC package with a different arrangement of address and data lines, for example in the first case 12 lines for address and 2 for data (1 input, 1 output) while in the second only 10 address lines but either 8 or 4 for data (4 if bi-directional lines are used). The pin configuration of a typical 1K $\times$ 4 memory chip is shown in Fig. 2.22. A_0 to A_9 are the address lines, D_0 to D_3 the (bi-directional) data inputs and outputs, and $\overline{WE}$ (*write enable*) selects the reading or writing of data.

Each memory chip in a system represents a part of the address space. The *chip select* input $\overline{CS}$ is used to enable the data input and outputs. Suitable *address decoding logic* would arrange their values to be set appropriately whenever the address of a word within its address space is generated by the CPU. Inputs with a 'bar' over their symbols are called *active-LO* since it is a low voltage which causes the required action (the normal state of the input would be HI, or high voltage). If an input has no 'bar' it is *active-HI* (its normal, inactive state being LO).

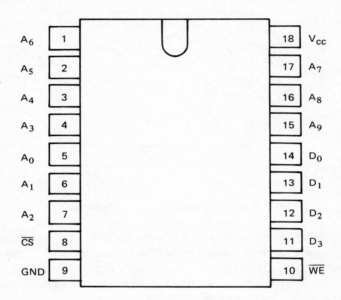

Fig. 2.22 – Pin configuration for a typical $1K \times 4$ memory IC.

A variety of types of memory is available. They are briefly described as follows:

Random access memory (RAM) should really be called read-write memory since it is the type which allows cells to be read from or written to. This is the type of memory used most commonly in computer systems for programs and data. The name RAM should apply also to the older technology magnetic core memory, which also allows reading and writing, but has come to be associated only with semiconductor memories. RAM is *volatile*, meaning that it loses its stored information should the electrical supply be switched off. In some computer applications loss of information cannot be tolerated, so their memory is equipped with battery back-up in case of an accidental loss of power.

Read-only memory (ROM), as the name implies, cannot be written to. This type of memory is used in systems where fixed programs or read-only data are required. ROMs are *non-volatile*. Most microcomputer systems have a minimum of one ROM area in their address space, containing a small fixed program called the *monitor* which allows the progress of programs in RAM to be controlled. *Bootstrap* programs which will be obeyed immediately when a computer is switched on can also be implemented in ROM. Various sub-divisions of ROM exist, distinguished by the way in which the information in the ROM is put there.

Mask-programmable ROMs are programmed by the manufacturer. A mask representing a pattern of 1s and 0s is used in the production process to set up the electrical connections which determine the ROM contents. The mask pattern is specified by the customer in advance. There is a masking charge which is such that it may be uneconomical to buy small quantities of mask-programmed ROMs.

Programmable ROMs (PROMs) are more accurately named field-programmable devices. They can be programmed by the user to his own requirements. Each memory cell is like a tiny fuse wire which when intact represents a 1. The 0 state is produced by passing a heavy electric current which breaks the fuse wire. A suitable *PROM programmer* device is required for this purpose: these can either be purchased or fairly easily built since the manufacturer specifies all the appropriate details for programming. The process is irreversible, so the user must be certain that the bit pattern is correct first time.

Erasable PROMs (EPROMs) are PROMs which can not only be user-programmed but erased as well. Mistakes either in the bit pattern or in programming the device can thereby be rectified (but note that the entire memory array is erased). Typically EPROMs can be erased and re-programmed up to 100 times. Erasing is possible by means of a perspex window in the top of the IC package, through which the memory chip itself is visible: the chip is exposed to ultraviolet (UV) light of a specified frequency for a recommended length of time. Note that when EPROMs are programmed and being used in a memory system the perspex window should be covered by an opaque material (such as black adhesive tape) to prevent any possible loss of information due to the UV in natural light.

Static RAMs are memory chips in which the cells do not lose information as long as electric power is supplied to them.

Dynamic RAMs, in contrast, need to be *refreshed* periodically otherwise the information would be lost. The distinction between the two types really applies only to MOS RAMs, since the bipolar memory chips are essentially static. The advantages of dynamic RAMs consist of the smallness of their memory cells compared with those in static memories, thus making possible the production of larger-capacity chips. Also they consume less power since when not selected (that is, with chip enable inputs disabled) they automatically go into a low-power-consumption *standby* mode. On the other hand dynamic RAMs do require external refresh circuitry which adds to the total memory cost. Generally, static RAMs are more cost-effective in small-memory systems (up to about 4K words). Otherwise dynamic RAMs are better.

Logic ICs include all digital chips other than memories and interfaces. The name is rather general and is intended to cover all logic devices which may be used to implement the CPU of a computer system, including gates and flip-flops at the SSI level, MSI devices such as registers and adders, and also microprocessor (LSI) units, themselves complete CPUs. Though logic devices form the largest category, little need be said about them in this section: they are available in IC packages, as already described. Chapters 3, 4 and 5, however, are primarily concerned with the design and implementation of logic within CPUs, where this category of IC will receive further attention.

Positive and negative logic

One last topic requires to be discussed in the present section. It brings the two streams, abstract and physical, back together. Basically, the point is that there are two forms of logic which may be used, depending on the way 1s and 0s are assigned to the physical realisations of HI and LO voltages. Either assignment is equally valid but the choice must be applied consistently. As we shall now see, the two forms of logic may be used to link together circuit design, expressed in terms of AND, OR and NOR operations, and implementation using the readily-available NAND gates. The following also makes additional points about logic symbols and their use.

When we buy a logic gate, the function which it performs must be specified by the manufacturer, but since the gate physically operates by voltage levels this function is specified in terms of voltages — for example a NOR gate has the following truth table:

A	B	$\overline{A+B} = F$
LO	LO	HI
LO	HI	LO
HI	LO	LO
HI	HI	LO

where the two states or signals are represented by a HI and LO voltage level (which will be +3.5V and 0V for TTL gates). Normally the manufacturer also specifies that the gate is a *positive-NOR* which means that by convention we assign 0 to LO voltage and 1 to HI. Then the truth tables will have the same functional appearance:

	A	B	F
	0	0	1
	0	1	0
(again, a NOR gate)	1	0	0
	1	1	0

This is called the *positive logic* convention.

If, however, we choose to assign 0 to HI and 1 to LO the truth table becomes:

A	B	F
1	1	0
1	0	1
0	1	1
0	0	1

which looks like a NAND gate in this *negative logic* convention.

With the same physical gate, then, two different functions can be realised depending on the logic convention chosen.

How do we indicate on a drawing the logic convention chosen? We do this by means of a small circle on the inputs or outputs of a gate. Since there are distinctive shapes for only AND and OR (not NAND, NOR) we therefore have to indicate the positive and negative versions of nominal gate functions as shown in Fig. 2.23.

Nominal Gate Function	Positive Logic		Negative Logic	
	Function	Symbol	Function	Symbol
AND	AND	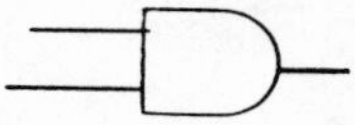	OR	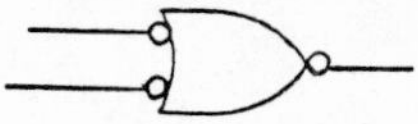
OR	OR	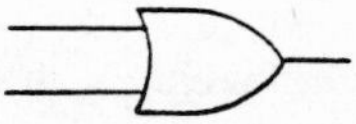	AND	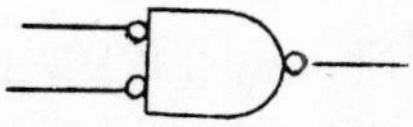
NAND	NAND	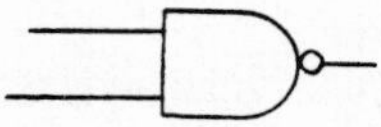	NOR	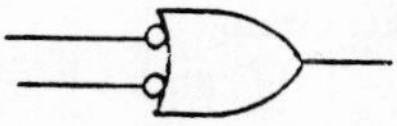
NOR	NOR	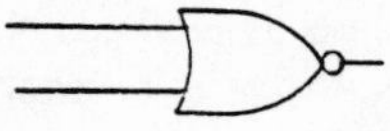	NAND	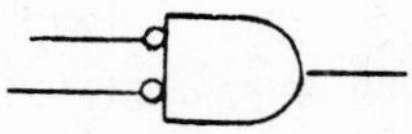
NOT	NOT	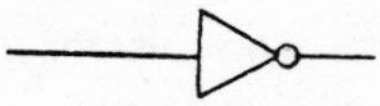	NOT	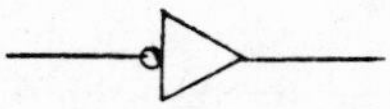

Fig. 2.23 – Positive and negative logic symbols.

To explain the table, let us look at the nominal AND function (in terms of HI and LO). If we apply negative logic, the truth table becomes that of an OR function (in terms of 1s and 0s). To indicate that negative logic is being used, the inputs and outputs are circled. Another way of looking at this change of logic convention is that it is equivalent to *inverting* the signal value. So we could say that if the inputs of an AND gate are inverted before the function and the output inverted after the function then the function looks like OR.

The reason why the NAND gate is shown and the AND symbol followed by a circle is that changing the logic convention at the output of an AND gate gives a NAND function, which is equivalent to inverting the output. The small circle can be interpreted as either a change of logic convention or as an inversion of the signal value.

Returning to the NAND gate with negative logic applied at inputs and output, this looks like the NOR function, which is normally written as

that is OR with an inversion at the output. The symbol, therefore, for a *negative* logic NOR is written as

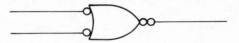

and since two inversions cancel each other out the symbol is actually written as

as indicated in the diagram. This symbol can be interpreted as a NOR working in negative logic, or as an OR gate with the inputs inverted or having undergone a logic change.

This can be used to maintain, in a logic diagram, the functions of AND and OR which arise naturally from truth tables and Boolean expressions, while enabling us to interpret circuit symbols as physical NAND (or NOR) gates.

For example, consider the exclusive-OR function $\overline{A}B + A\overline{B}$. The circuit diagram can be drawn as in Fig. 2.24(a).

It would be helpful to keep the symbols for AND, OR so that we easily recognise the function which the circuit performs. By changing the logic convention between gates we can do this, as Fig. 2.24(b) shows. The gates keep their original functions but physically each can be realised by a 2-input NAND gate.

Notice that the NOT gate has a positive and negative logic version — it can be considered as an amplifier ——▷—— with *either* a logic convention change or inversion at input (negative logic) *or* output (positive logic).

The circuit in Fig. 2.24(b) can be realised entirely by 2-input NAND gates,

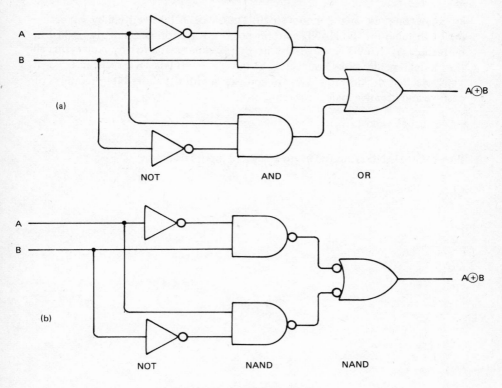

Fig. 2.24 – Exclusive-OR logic circuit:
   (a) in terms of AND, OR, NOT gates.
   (b) in terms of NAND, NOT gates.

since it is very easy to implement NOT using NAND. This can be done in two ways, either

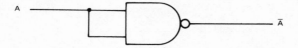

by connecting both inputs together, or

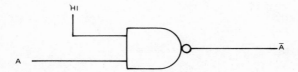

by connecting one of the inputs to HI. These can both be verified by inspecting
the truth table for the NAND gate (in positive logic form). Similarly, NOT can
be realised by NAND gates with more than 2 inputs — either by connecting all
the inputs together or by connecting all but one to HI (via $1K\Omega$ resistors).

Note that de Morgan's laws are consistent with the above interpretations of
positive and negative logic, for example

$$\bar{A} + \bar{B} = \overline{AB}$$

shows that NAND is equivalent to a negated-input OR gate.

CHAPTER 3

Combinational and sequential logic

3.1 REPRESENTATION OF LOGIC CIRCUITS

The terms *combinational* and *sequential* have already been defined for logic elements in Section 2.1: the outputs of a sequential element depend on previous as well as present inputs, whereas combinational elements are a special case in which only the present inputs affect their output values. From now on the word 'element' will be replaced by 'circuit' since any circuit can be represented as a black-box element.

As shown in Fig. 3.1 sequential circuits may be further sub-divided into the *asynchronous* and *synchronous* classes.

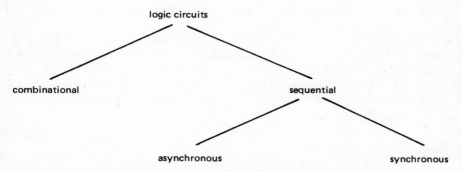

Fig. 3.1 – Classification of logic circuits.

In Sections 2 and 3 of this chapter combinational circuits will be studied first, followed by a discussion on sequential logic. Although asynchronous circuits are mentioned, and an example given, it is the synchronous type which is most frequently used in computers and consequently they will receive substantially more treatment. Design techniques and example circuits will be presented for both combinational and synchronous sequential logic.

In the present section we concentrate on the various different representations of logic circuits, some useful in the design techniques presented later while others are simply ways of presenting circuits to make their functions clearer. In general, logic circuits can be represented by the following means:

> written descriptions
> mathematical expressions
> tables/diagrams

and quite commonly by all three together. Tables or diagrams (for example *truth tables*) are usually very effective in communicating the meaning of (complex) ideas but often require annotation, particularly to explain symbols which may have been used from necessity. The compact, expressive power of mathematics (in the shape of Boolean algebra) may also be required to add precision to the (often generalised) table or diagram.

The combinational logic gates were expressed initially in this way — truth tables, written explanations and Boolean expressions. The ideas involved in these gates are so simple that we can replace the three modes of expression by a single distinctively-shaped symbol. There are also fortunately few different types of combinational gate and correspondingly few symbols to remember. The flip-flop varieties will also be introduced by means of tables, diagrams and a written explanation (in Section 3.3). They, too, are few and sufficiently simple to be represented by rectangular shapes bearing letters to label inputs and outputs. The input identities are adequate reminders of the function which the flip-flop performs.

MSI and LSI circuits of greater complexity than gates and flip-flops are represented also by simple rectangular shapes, in these cases showing the pin configuration of the IC packages as described in Section 2.4. All the pins are labelled, but the complexity of the circuits is such that a suitable explanation needs to be attached. In effect these circuits are represented by their respective data sheets.

Some pictorial representations

It is sometimes helpful to describe combinational logic circuits using pictorial forms related to the physical and abstract sides of digital logic.

Switching representations can be made as illustrated in Fig. 3.2. Each switch is equivalent to a Boolean variable, and their interconnections to the form of the Boolean expression.

The basic three Boolean operations AND, OR and NOT are shown. The AND and OR operations are combinations of switches connected in series and parallel respectively. Inversion is achieved by a double switch arrangement: when one is closed, the other is open and vice versa. The value 0 is represented by an open, 1 by a closed switch. These switch forms may be employed in

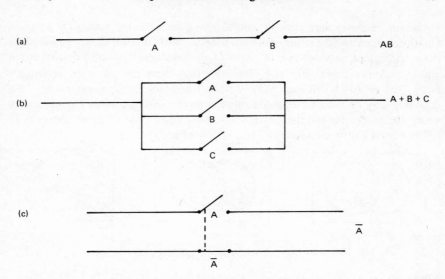

Fig. 3.2 — Switching representation of (a) AND, (b) OR, (c) NOT.

representing more complex combinational circuits or more useful still in demonstrating the postulates and theorems of Boolean algebra. For example Theorem T5:

$$A + AB = A \text{ (see Section 2.2)}$$

follows from Fig. 3.3 without further explanation.

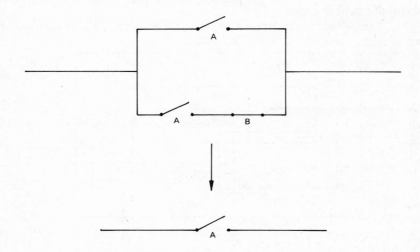

Fig. 3.3 — Demonstration of theorem T5: $A + AB = A$.

Venn diagrams may also be used to give a pictorial representation of combinational circuits. These diagrams are used in mathematical theory to illustrate the relationships between sets in terms of the set operators: intersection, union and complementation. Fig. 3.4 illustrates that these operators are essentially equivalent to AND, OR and NOT. Just as for switching representations, Venn diagrams may be used to illustrate more complex combinational circuits and to verify the postulates and theorems of Boolean algebra. For example, theorem T5 follows from a little consideration of Fig. 3.4(a) and (b).

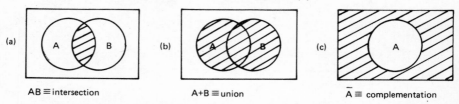

Fig. 3.4 – Venn diagram representation of (a) AND, (b) OR, (c) NOT.

Karnaugh maps

Karnaugh maps, or *K-maps* as they will be called hereafter, are very useful in practice for representing combinational circuits. Their main purpose, as will be explained in Section 3.2, is to aid the process of *minimisation* whereby logic circuits are implemented as efficiently as possible (as we shall see there are various criteria by which minimisation may be measured).

The K-map is really an extension of the combined ideas of Venn diagrams and truth tables: it combines the idea of representing all possible combinations of *n* Boolean variables on a plane, with the idea of listing all the variable values and output results in a table.

The Venn diagram can be used to show, for example, all combinations of three variables A, B and C as in Fig. 3.5.

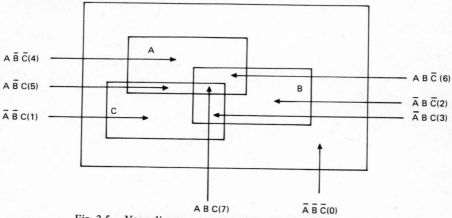

Fig. 3.5 – Venn diagram representation of 3 variables.

We can give each of the expressions a numerical value as shown in brackets. The value corresponds to the binary equivalent if we use 1 and 0 instead of A and $\bar{A}$ and so on. A particular logical expression can be plotted on the diagram by shading the area or areas desired, or by assigning an output value of 1 to the shaded areas and 0 to the unshaded. Thus, for example, the AND function for 3 variables is represented by shading area number 7, or by placing a 1 there (and 0s elsewhere).

Notice that in moving horizontally or vertically from one sub-region to another there is a change in only one variable (or a change in only one binary digit). Karnaugh maps are based on the same principle, except that the diagram takes the form of a map as in Fig. 3.6(a). The squares in the map have been given the same numerical values as in the Venn diagram.

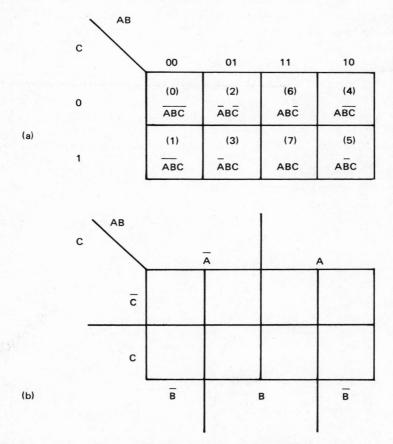

Fig. 3.6 — K-map representation of 3 variables: (a) significance of each square. (b) division into regions.

Notice that *reflected binary* (or *Gray code*) is used for listing the input variable values along the top and sides of the map — because Gray code involves a change in only one binary digit from one number to the next, as illustrated in Fig. 3.7. This figure shows the pure binary codes for the decimal digits 0–9 and their corresponding Gray code values. It is also called reflected binary because of the reflection symmetry which exists about a horizontal line drawn after two, four, eight rows and so on. For example this symmetry is evident for the least significant two bits from the line under the fourth row in the Gray code column.

Fig. 3.6(b) demonstrates the division of the K-map into regions according to the values of the individual variables. This can be useful for the minimisation procedure to be described in the next section.

K-maps may be used to represent 2-, 3- or 4-variable expressions on a single map. For 5 variables two adjacent maps are required; for 6, four maps. The K-map is impracticable for any greater number of variables. Outlines for 2-, 4-, 5- and 6-variable maps are given in Fig. 3.8, divided into regions as in Fig. 3.6(b).

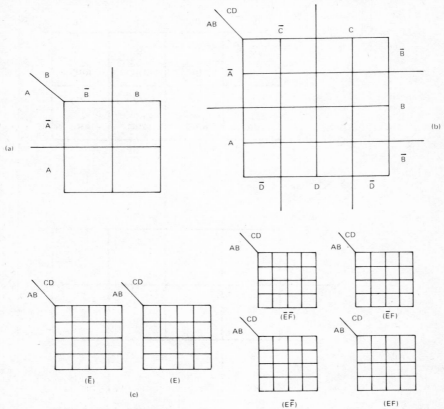

Fig. 3.8 — K-map outlines for (a) 2, (b) 4, (c) 5, (d) 6 variables.

Care must be taken in noticing which variables are presented horizontally and which vertically. For example, Fig. 3.8(b) is differently organised from Fig. 3.6(b) in this respect.

Decimal	Pure binary	Gray code	
0	0000	0000	
1	0001	0001	
2	0010	0011	↑
3	0011	0010	
4	0100	0110	↓
5	0101	0111	
6	0110	0101	
7	0111	0100	
8	1000	1100	
9	1001	1101	

Fig. 3.7 – Correspondence between decimal, pure binary and Gray code.

State tables, state diagrams and finite-state machines

All the above representations apply to the special case of combinational circuits. Sequential logic systems, otherwise known as finite-state machines (FSMs) as explained in Section 2.1, have a more complex structure and necessarily require to be represented in a more suitable form.

The structure of FSMs may be elaborated as follows: an FSM consists of five sets (I, O, S, T_s, T_o) where

$$I = \text{input symbols}$$
$$O = \text{output symbols}$$
$$S = \text{internal states}$$
$$T_s = \text{inter-state transistions}$$
$$T_o = \text{output transitions}$$

Less formally, any FSM has a finite number of states, in any one of which it can be at a given time. It changes from one state to another in discrete timesteps, depending on the input symbols and on the inter-state transitions. The FSM also outputs symbols as it progresses from state to state according to the output transitions. Thus, given any state and input symbol, the movement of the FSM is entirely predictable. This is the same as the generalised black box behaviour. The FSM can be represented in either of two ways, by a state table or a state diagram.

A *state table* is an extension of a truth table which includes extra columns for present and next internal states. The general form of a state table is shown in Fig. 3.9(a), and a very simple example table in (b).

	Present state/Inputs	Next state	Outputs
(a)	all combinations of present state and input symbols	next state for each combination of present state and inputs	output symbol for each combination of present state and inputs

	S	I	S'	O
	0	0	0	0
(b)	0	1	1	1
	1	0	0	0
	1	1	1	1

Fig. 3.9 − State table: (a) general form, (b) example.

For the simplest sequential circuits there may be no need to distinguish between next state S' and output O − as in Fig. 3.9(b) − thus a reduced version of the state table may be used.

State diagrams contain exactly the same information as the tabular form. Internal states (or nodes) are represented by circles, and inter-state transitions by lines (called arcs) joining the circles. The lines are further labelled with the input symbol (or symbols) leading to this transition, and with the resulting output symbol. Fig. 3.10 gives the state diagram version of the example in Fig. 3.9(b).

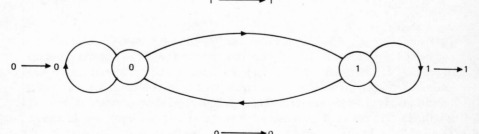

Fig. 3.10 − Example of a state diagram.

The Moore/Mealy model

A very important representation of sequential circuits, as shown in Fig. 3.11, is due to E. F. Moore and G. H. Mealy. This gives a view of a sequential circuit in terms of the elements with which it will be constructed: a combinational logic network and a memory by which previous inputs are remembered. It also shows the element of *feedback* which is fundamental to the construction of sequential circuits.

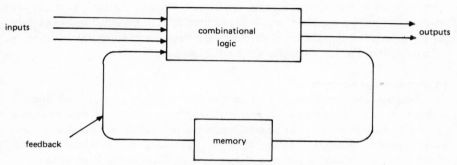

Fig. 3.11 – Moore/Mealy model of a sequential logic circuit.

Moore and Mealy independently reported on the structure and synthesis of sequential circuits but their models, although different in detail, are both essentially as shown in Fig. 3.11. Their work followed on from previous investigations into synthesis procedures by D. A. Huffman, who used a form of state table (called a flow table) to represent circuits. Huffman and Moore are credited with introducing the notion of an internal state into sequential design procedures (i.e. equating sequential circuits with FSMs). Moore, in addition, developed the idea of a state diagram for starting off the design process.

3.2 COMBINATIONAL LOGIC DESIGN

We begin the study of logic circuit design by considering combinational logic problems. This class of problem is characterised by the fact that output values are a function of only the present inputs. Furthermore, for the moment it can be assumed that a change of input values has immediate effect on the outputs.

In computer systems, there are many combinational circuits, such as

> adders
> ALUs
> carry look-ahead generators
> comparators
> code converters
> $n \times m$ decoders

priority encoders
multiplexers/demultiplexers
parity generators/checkers

These are all circuits available in MSI form. Within the CPU of a typical computer they would be amalgamated with registers to give a larger functional unit. Registers, being sequential circuits, will be studied in the following section. Our purpose in the present section, however, is twofold: to show how circuits such as the ones listed above are designed from SSI components, and to explain in the process of demonstrating examples the structure of some important parts of a computer. Further examples will be discussed in Chapter 4.

A simple design example

To illustrate the design process let us consider a simple problem not directly related to computer systems. It is a switching problem — how to construct a circuit to enable one light to be equipped with two switches. This arrangement is common in domestic lighting schemes. Expressing the problem in detail, there are two switches A and B, and one light L. When A and B are both down or both up L is off, otherwise L is on.

This statement is more concisely expressed in a truth table:

A	B	L
down	down	off
down	up	on
up	down	on
up	up	off

Let us substitute 0 for switch position down and light off, and 1 for up and on. We now have

A	B	L
0	0	0
0	1	1
1	0	1
1	1	0

As described in Section 2.2, the output L can be expressed in terms of A,B and the fundamental operators AND, OR and NOT. Consider the rows only where an output of 1 appears. It is those combinations of inputs which lead to the Boolean sum-of-products expression:

$$L = \bar{A}B + A\bar{B}$$

From this point we are in a position to draw a circuit diagram, depending on the availability of components with which to implement the circuit.

A suitable switching circuit can be drawn as follows:

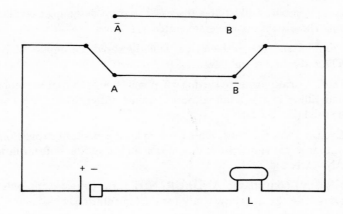

Notice that the function shown in the truth table is the exclusive-OR (EOR), as defined in Section 2.1. Although EOR is itself available as a (quad) MSI package it is instructive to show its implementation using SSI gates. Suitable circuits were presented in Fig. 2.24(a) and (b), the first using AND, OR, NOT and the second the corresponding NOT, NAND implementation. These circuits can be drawn from the Boolean expression for L.

Fig. 2.24(b) can further be converted into a circuit using 2-input NAND operations only. This is shown in Fig. 3.12.

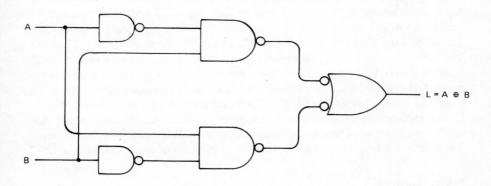

Fig. 3.12 – NAND gate version of the exclusive-OR (EOR) function.

As described in Section 2.4, NOT may be implemented using NAND either by connecting all inputs together or by connecting all but one to HI. Either way, the NAND version of NOT may be indicated by a single-input NAND gate, as in the diagram.

This simple problem illustrates the basic steps in designing a combinational logic circuit. These steps may be enumerated as follows:

(1) Think round the problem, particularly about the inputs and outputs which the circuit will have.

(2) Draw a truth table by listing all possible input variable combinations and filling in the required output values. Alternatively a K-map could be used (but see later).

(3) Extract from the truth table (or K-map) the Boolean expression for the output value and write it down as a sum-of-products form in terms of AND, OR and NOT.

(4) (Usually) convert the AND, OR, NOT form to one using only NAND gates. The circuit is now ready for implementation.

The simplicity of the lightswitch problem is such that little thought need be put into step (1). For more complex, multiple-input, multiple-output problems this step may be significantly more difficult. In the case of multiple outputs a truth table is required (step (2)) for each output. Alternatively all the truth tables can be merged as long as the outputs are all functions of the same set of inputs.

In general, the extracted Boolean expression will be much more complex than that for L above. It is often possible to reduce the circuit in terms of the number of gates (or IC packages) used, the number of interconnections between gates and the *levels of logic* in the circuit. Referring to Fig. 3.12, there are three levels in the circuit, corresponding to the maximum number of gates which an input signal passes through on its way to the output. All combinational circuits can be implemented, in theory, using only three levels of logic because they can all be expressed in a sum-of-products form with the three levels NOT, AND and OR. In practice the expression may have to be manipulated or factorised because the available gates do not have sufficient inputs for direct implementation of the three-level sum-of-products form. Such factorisation results in an increased number of levels of logic. The importance of minimising levels of logic is that each level corresponds to a gate delay in the circuit, in other words the smaller the number of levels, the faster the circuit will operate. The next example will illustrate how a circuit may be minimised in terms of the above criteria.

Wiring diagrams
As mentioned in Section 2.4 circuits are implemented using IC packages containing the appropriate logic elements. Implementation is the process of translating

the circuit diagram into an interconnected array of .ICs (or IC sockets) on a circuit board. For this purpose the wiring diagram is important.

The first step in implementing a logic circuit is therefore to add point-to-point wiring information to the designer's circuit. Alternatively a wiring list may be made up separately, listing the pins in each IC and the other pins/ICs they are to be connected to. In either case a layout of the circuit board must first be planned, to show the positions of ICs, as in Fig. 3.13.

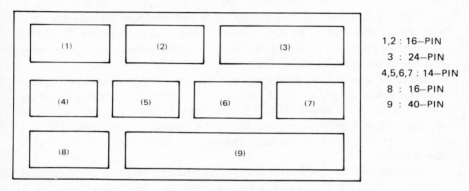

Fig. 3.13 – Example of an IC board layout.

Preferably the IC positions should be chosen to minimise the amount of wire used in making the interconnections. For all but the smallest of circuits this is a difficult task. Sometimes the criterion is to use the smallest board possible. This can be achieved by arranging the packages so that they occupy the minimum area, depending on their sizes. Fig. 3.13 shows an arrangement chosen with this aim in mind. Generally it will not be possible to satisfy this criterion as well as minimising the amount of wire used.

If the wiring details are to be added to the circuit diagram the IC numbers on the board layout have to be used to label the appropriate symbols on the diagram as in Fig. 3.14, which illustrates one section of a circuit.

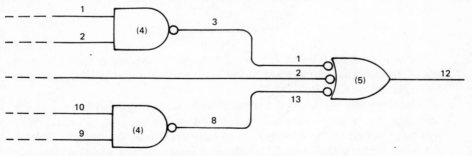

Fig. 3.14 – Section of a circuit diagram showing wiring connections.

Of the three logic gates two are to be implemented using a type 7400 IC, or quad 2-input NAND gate package, labelled as (4). The third, a 7410 or triple 3-input NAND, corresponds to IC number (5). Implementing the interconnections is a matter of reading the pin/package details directly from the drawing, and wire-wrapping each output pin to the appropriate input.

Design example — the binary adder

The basis of the arithmetic and logic unit (ALU) in computers is the *binary adder*. This takes two binary digit operands and produces their binary sum. The adder is also used to subtract numbers by the 2s complement (or the 1s complement) method and may be employed in multiplication and division units which are based on repetitive methods.

When we add two numbers manually, whether they be decimal or binary, we start at the least significant end and add the two corresponding digits together — we write down the result of the addition and take the carry digit, if any, to the next-to-least significant position where the procedure is repeated. This is a serial process; that is, we use the same 'circuit' (our mental faculties) to add each stage in turn.

Similarly, we can perform serial addition using electronic circuitry, providing one circuit for adding two digits and presenting the operands one pair at a time. We can represent this system as follows:

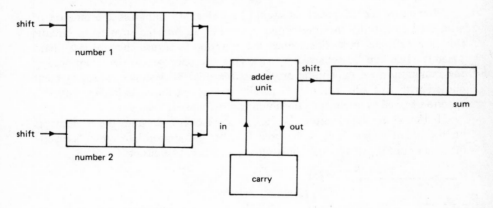

A more costly but effective method is to provide one adder circuit for each pair of digits in the numbers and to perform the addition in parallel, that is all the stages together.

Unfortunately the addition cannot truly be done in parallel because of the carry digit propagation delay — it takes a finite time for each successive carry digit to be passed to the next stage. This delay determines the speed with which two n-bit numbers are added. A method of improving the addition time will be presented later.

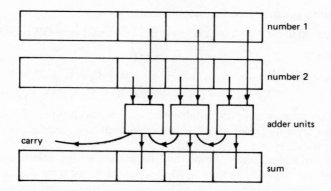

We now design a single-stage adder unit using combinational logic. It is usual to distinguish between the half-adder and the full-adder circuits, which are shown diagramatically below.

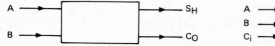 

In both cases, A and B are the binary operands, S is the resulting sum and C_O is the carry-out bit. The difference between the two is that the half-adder neglects the carry-in C_i from the previous stage while the full-adder takes it into account.

Truth-tables for these circuits can easily be constructed:

A	B	S_H	C_O
0	0	0	0
0	1	1	0
1	0	1	0
1	1	0	1

C_i	A	B	S_F	C_O
0	0	0	0	0
0	0	1	1	0
0	1	0	1	0
0	1	1	0	1
1	0	0	1	0
1	0	1	0	1
1	1	0	0	1
1	1	1	1	1

half-adder   full-adder

The Boolean equations for the outputs are:

HALF-ADDER $S_H = \bar{A} B + A \bar{B}$ (the exclusive-OR function)

$C_O = AB$

FULL-ADDER $S_F = \bar{C}_i \bar{A} B + \bar{C}_i A \bar{B} + C_i \bar{A} \bar{B} + C_i A B$

$C_O = \bar{C}_i A B + C_i \bar{A} B + C_i A \bar{B} + C_i A B$

Notice that amalgamated truth tables are used for both types of adder. Alternatively two separate tables could be drawn in each case, one for the sum and the other for the carry-out.

The half-adder is very simple and its corresponding circuit consists of an EOR part (for the sum) and the AND function of A and B for the carry-out.

On the other hand, the full-adder equations for S_F and C_O will both produce circuits consisting of four 3-input gates (for the product terms), one 4-input gate (for the sum) and three inverters (for C_i, A and B). Quite often it is possible to reduce the size of such circuits because of their inherent redundancy: variables can be eliminated from the expressions to make the product terms both smaller (i.e. fewer variables in each) and fewer in number. This can be done in one of two ways — using Boolean algebra to manipulate the expressions, or by plotting the expressions on K-maps.

The sum and carry expressions for the full-adder can be re-written using *algebraic manipulation* as follows:

$$S_F = C_i (AB + \bar{A} \bar{B}) + \bar{C}_i (\bar{A} B + A \bar{B})$$
$$= C_i \bar{S}_H + \bar{C}_i S_H$$

(since by applying de Morgan's laws $\overline{A B + \bar{A} \bar{B}} = \bar{A}B + A \bar{B}$)

Thus another way of implementing the full-adder sum is to use two half-adder circuits (i.e. two EOR gates): Hence the names. The expression for S_F cannot otherwise (in terms of AND, OR, NOT) be reduced in size. However, the carry can be written as

$$C_O = \bar{C}_i A B + C_i A B + C_i \bar{A} B + C_i A B + C_i A \bar{B} + C_i A B$$

(adding two redundant $C_i AB$ terms has no effect on the value of the expression)

$$= C_i A (B + \bar{B}) + C_i B (A + \bar{A}) + AB (C_i + \bar{C}_i)$$
$$= C_i A + C_i B + A B$$

(since $A + \bar{A} = 1$ and $A.1 = A$).

Thus we have a simplified form for C_O, with three 2-variable terms instead of four each of three variables. With the redundancy removed we can apparently now draw a minimal circuit. However, by manipulating the C_O expression in a different way:

$$C_O = A B (C_i + \bar{C}_i) + (C_i \bar{A} B + C_i A \bar{B})$$
$$= A B + C_i S_H$$

we can make use of the expression for S_H already available from the sum part of the adder. This will produce a smaller circuit than the previous manipulation would have. The circuit for a full-adder based on half-adders (EOR gates) is drawn in Fig. 3.15. The AND, OR circuit for the carry has been drawn directly in NAND gate form.

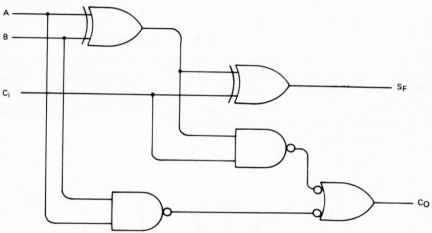

Fig. 3.15 – Binary full-adder circuit diagram.

Reducing the size of expressions using Boolean algebra is basically simple: just look for the possibility of factorising the expression such that a term of the form $(A + \bar{A})$ emerges, which reduces to 1. In this way redundant variables are eliminated. If possible expressions should be closely inspected to see if they are amenable to manipulation to take advantage of common sub-expressions (such as S_H above), as this is likely to produce a smaller overall circuit.

A more convenient way to eliminate redundant variables is by use of *K-maps*. This method allows elimination to be effected with a quick visual scan because terms containing potentially redundant variables will be adjacent on the map. Consider the K-map for S_F:

C_i \ AB	00	01	11	10
0		$\bar{C}_i\bar{A}B$		$\bar{C}_i A\bar{B}$
1	$C_i\bar{A}\bar{B}$		$C_i AB$	

S_F

which is more usually drawn as:

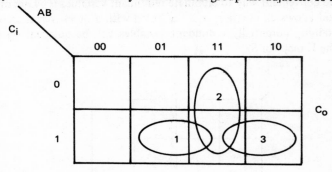

because the value of each expression is obvious from the position of the plotted 1s. Terms occupying adjacent squares horizontally or vertically differ only in the value of one variable: they may be combined to eliminate that variable according to the rule $A + \bar{A} = 1$. In the case of S_F there are no adjacent squares occupied by 1s therefore the expression cannot be reduced.

For C_O the K-map is:

There are adjacent squares containing 1s — two pairs horizontally and one pair vertically. These are indicated by circles which group the adjacent 1s together:

Group 1 consists of $C_i\bar{A}B$ and C_iAB which reduces to C_iB (eliminating A). Groups 2 and 3 similarly reduce to AB and C_iA, eliminating C_i and B respectively. Thus C_O may be written:

$$C_O = C_iA + C_iB + AB$$

the same expression which was obtained by the first of the two algebraic manipulations. The K-map inspection does allow a rapid elimination of any redundant variables and as such is very useful for reducing the size of circuits. It does not help the process of manipulating expressions to make use of already available sub-expressions. We shall be returning to the topic of minimisation criteria.

The use of K-maps

The adder example showed briefly how K-maps can be used to represent and reduce the size of combinational circuits. Let us more fully explore their use in logic design.

K-maps are really just a rearrangement of the information in a truth table. The 1s in the output column of the table are plotted on the squares of the K-map, and the 0s are represented by blank squares. The position of the squares is arranged so that if each one represents a combination (product term) of n variables, then $n - 1$ variables have the same value in adjacent squares while one variable is complemented in one square but not in another. Adjacent squares for a 3-variable map are shown in Fig. 3.16 by double-headed arrows. Note that adjacency includes wrapping round the map horizontally or vertically from each square at an edge to the square at the opposite edge.

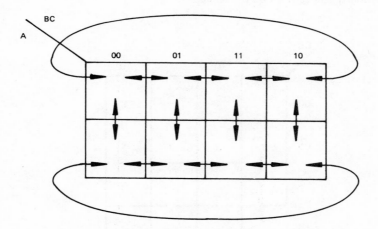

Fig. 3.16 – Adjacent squares on a 4-variable K-map.

Adjacency on a 4-variable map is similar, though for maps with 4 or more variables it is more difficult to spot adjacent 1s. For 5-variable maps (two 4-variable maps side by side) it helps to imagine one of the two maps in a plane above the other so that adjacent squares in the fifth variable are aligned along the third dimension.

In combinational logic design, K-maps may usefully be employed in a stage following the formation of the truth table. The expression is plotted on the map, reduced in size if possible and finally rewritten. Alternatively the K-map may be used instead of the truth table as a direct means of plotting the desired logical function, and only the reduced form of Boolean expression written down. Expressions in a sum-of-products form derived from truth tables are suitable for direct translation onto a K-map: one product term maps onto one square.

Sometimes it may be wished to reduce the size of an expression, if possible, such as the following:

$$F = \bar{A}\bar{B}\bar{C} + \bar{A}B + \bar{A}BC\bar{D} + AC\bar{D}$$

This is a 4-variable expression (variables A, B, C, D) but is not in *canonical* form, that is each product term does not contain all four variables. For direct mapping of each product onto a square of the K-map the expression should be converted to canonical form. This is done by expanding those terms not containing all four variables by ANDing them with the sum of each missing variable and its complement (since $A + \bar{A} = 1$ this has no effect on the value of the expression). Using the example:

$$F = \bar{A}\bar{B}\bar{C}\,(D+\bar{D}) + \bar{A}B\,(C+\bar{C})\,(D+\bar{D}) + \bar{A}BC\bar{D} + AC\bar{D}\,(B+\bar{B})$$
$$= \bar{A}\bar{B}\bar{C}D + \bar{A}\bar{B}\bar{C}\bar{D} + \bar{A}BCD + \text{-----------} \quad \text{(expanded)}$$

The function is now plotted on the K-map:

AB \ CD	00	01	11	10
00	1	1		1
01	1	1	1	1
11				1
10				1

F

Product terms in a canonical sum-of-products expression each occupy one square, the smallest area on a map, and are thus sometimes called *minterms*. K-maps are also sometimes referred to as minterm maps.

How do we group squares in general to give a reduced expression? The rules are that squares should be combined in groups of 2, 4, 8 and so on in powers of 2. In the 4-variable case the groupings can be 2, 4 or 8 squares (thus eliminating 1, 2 and 3 variables respectively). The largest possible groupings should be sought first. The squares being combined are circled and the aim is to cover all the squares containing 1s in the most effective way.

Squares can belong to more than one circled group — they can be covered more than once. Circled groups are known as *prime implicants* (PIs) of the plotted function, and groups containing at least one square which cannot belong to any other group are known as *essential prime implicants*.

The prime implicants in the example are shown circled:

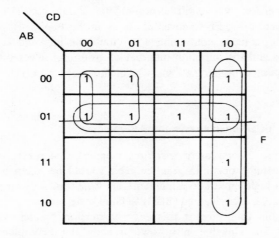

There are four circled groups, that is four PIs, each of four squares. However, the PI

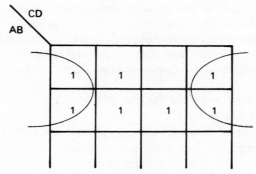

(corresponding to $\overline{A}\overline{D}$) is non-essential, containing only squares which belong to other circled groups. As such, this PI should not be included in the final expression. The other three PIs are all essential and between them cover all the 1s on the map. Thus the final reduced expression is

$$F = \overline{A}B + \overline{A}\overline{C} + C\overline{D}$$

To write down the reduced terms corresponding to each PI it is helpful to consider the K-map as a set of regions for each variable as in Fig. 3.6(b). The value of each PI is determined by the intersection it makes across the regions on the map.

In the above example the final reduced expression is unique since all 1s are covered by essential PIs. In some cases the essential PIs are not sufficient to cover all 1s, and there may be a choice of other PIs which cover the remaining 1s. The final expression, which will be non-unique, should consist of the essential PIs and sufficient other PIs to cover all the 1s on the map.

To summarise, squares containing a 1 should be included in the largest possible groupings and the minimum number of groupings selected to cover the function. Each square containing a 1 must be included at least once.

Design example — BCD to Gray code conversion

Some functions are *incompletely specified,* meaning that some combinations of input values cannot occur in practice. This should be taken into account in the design of a logic circuit to implement the function. An example is in the conversion of binary-coded-decimal (BCD) to Gray code values.

The truth table to represent this code conversion problem is essentially contained in Fig. 3.7. This does not, however, show the incompleteness of the function. BCD values require four bits to store the ten possible decimal values 0 to 9, leaving six unused combinations of the bits. These unused combinations cannot occur in practice. When the Boolean expressions corresponding to the code conversion are to be reduced (if possible) we can choose whether to include any of the outputs which would result from unused input combinations, if these will help the reduction process: these outputs are marked by Xs on the truth table and K-map to indicate *don't care conditions* (they can be 0 or 1 as we choose).

Fig. 3.17 shows the truth table for the BCD to Gray code conversion, including don't care conditions.

This is a multiple-output problem. Although all four inputs G_3, G_2, G_1, and G_0 are amalgamated into one truth table each requires a separate K-map for the reduction process. The K-map for the least significant output is as follows:

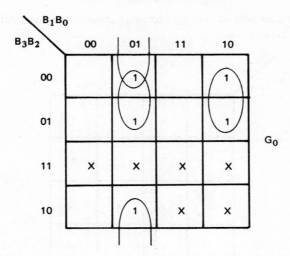

Without using the don't care conditions the grouped 1s give a final expression

$$G_0 = \bar{B}_3 B_1 \bar{B}_0 + \bar{B}_3 \bar{B}_1 B_0 + \bar{B}_2 \bar{B}_1 B_0$$

BCD				Gray code			
B_3	B_2	B_1	B_0	G_3	G_2	G_1	G_0
0	0	0	0	0	0	0	0
0	0	0	1	0	0	0	1
0	0	1	0	0	0	1	1
0	0	1	1	0	0	1	0
0	1	0	0	0	1	1	0
0	1	0	1	0	1	1	1
0	1	1	0	0	1	0	1
0	1	1	1	0	1	0	0
1	0	0	0	1	1	0	0
1	0	0	1	1	1	0	1
1	0	1	0	X	X	X	X
1	0	1	1	X	X	X	X
1	1	0	0	X	X	X	X
1	1	0	1	X	X	X	X
1	1	1	0	X	X	X	X
1	1	1	1	X	X	X	X

DON'T   CARE   CONDITIONS

Fig. 3.17 – BCD to Gray code conversion showing don't care conditions.

A quick visual scan tells us that by including appropriate don't care conditions (effectively as 1s) as shown:

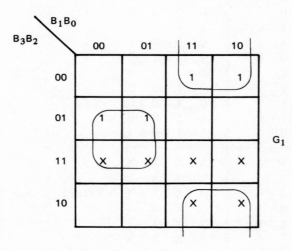

G_0

we obtain a very much reduced expression:

$$G_0 = B_1\bar{B}_0 + \bar{B}_1 B_0 = B_1 \oplus B_0$$

eliminating B_3 and B_2, and reducing the number of terms to only two. The K-maps and reduced expressions for the other outputs are shown below:

$$G_1 = B_2\overline{B_1} + \overline{B_2}B_1 = B_2 \oplus B_1$$

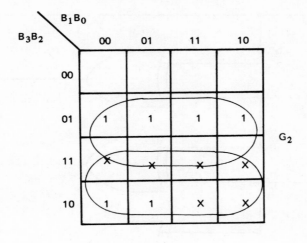

$$G_2 = B_3 + B_2$$

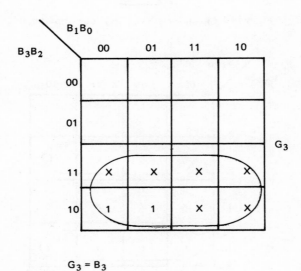

$$G_3 = B_3$$

In the cases of G_2 and G_3, groupings of eight squares are circled, thus eliminating three variables.

A circuit diagram for the BCD to Gray code conversion is illustrated in Fig. 3.18, using EOR gates. Although the expression above for G_2 has been reduced as much as possible it is better in this case not to use all the don't care conditions:

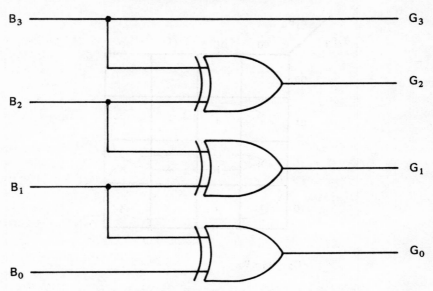

Fig. 3.18 – Circuit for BCD to Gray code converter.

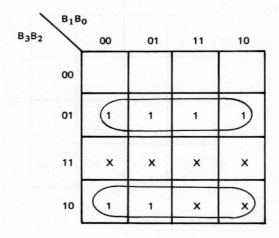

but to circle only the groups shown to give

$$G_2 = B_3\bar{B}_2 + \bar{B}_3B_2 = B_3 \oplus B_2$$

Since EOR gates are available four in one IC package (with designation 7486) the converter may thus be implemented with one IC (leaving one spare EOR gate in it).

Some notes on circuit minimisation

Using K-maps, or indeed Boolean algebra, logic expressions can be reduced to a minimum size by eliminating redundant variables. This reduction process is usually called *minimisation*. Generally speaking, minimisation consists of reducing the number and size of the product terms in a Boolean expression to the minimum, although in practice the most effective solution depends on specific circumstances, for example the availability of certain types of gate, or the existence of common sub-expressions within a large circuit.

In the first generation of computers, AND and OR were implemented by inexpensive diode logic, but the NOT operations required a vacuum tube circuit which was relatively expensive. One of the principal aims, therefore, in building combinational circuits was to keep the number of inverters to a minimum. For sequential circuits amplification had to be provided to implement the feedback path (see Fig. 3.11). Again this was expensive to provide, so the aim in these circuits was to minimise the number of feedback loops. Second generation gates provided both inversion and amplification so the early criteria disappeared, to be replaced by a general requirement to reduce the total number of gates.

More recently, the cheapness of gates in IC packages has altered the balance of the criteria. The cost of wiring, including both materials and labour, has risen so much in comparison with IC costs that it is often cheapest to minimise the number of interconnections between gates. Small circuit size is usually considered important: in these circumstances keeping down the IC package count and optimising their layout on the circuit board are key factors. Minimising the number of levels of logic in a circuit leads to faster circuit operation and may be the most important criterion in particular implementations.

Reducing the number of product terms in an expression using a K-map (or otherwise) corresponds to reducing the number of gates in the circuit, while reducing the size of each product is equivalent to minimising the interconnections. For example, consider the 4-variable expression introduced earlier:

$$\overline{A}\overline{B}\overline{C} + \overline{A}B + \overline{A}\overline{B}\overline{C}\overline{D} + AC\overline{D} \tag{3.1}$$

Using a K-map, we know that the expression reduces to

$$\overline{A}B + \overline{A}\overline{C} + C\overline{D} \tag{3.2}$$

which appears to be much cheaper to implement than (3.1).

The number of interconnections is important not only because of wiring costs but also because of the fan-in and fanout factors for the available gates. Fan-in is determined by the number of input pins for a gate — 2, 3, 4 or 8 for NAND gates. Depending on the number of inputs per gate a different number of gates can be accommodated in the IC package, as explained in Section 2.4. Fan out, the number of inputs which may be connected to a single gate output, is not physically limited by a pin count so care must be exercised by the designer in ensuring that gates are not overloaded by exceeding the permitted fanout. A fanout of 10 is usual for TTL gates, but such a figure must be carefully interpreted, since (usually) it is quoted for gates of the same type.

Continuing the example, the availability of only 2-input NAND gates means that (3.2) must be factorised to avoid the necessity of a 3-input gate. One of the ways this can be done is as follows

$$\bar{A}(B + \bar{C}) + C\bar{D} \qquad\qquad (3.3)$$

Circuit diagrams for all three versions of the expressions are given in Fig. 3.19.

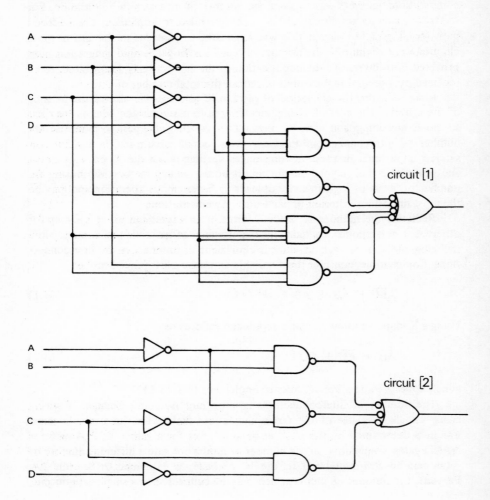

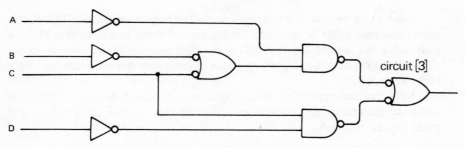

Fig. 3.19 – Three circuits to implement the same expression.

The gate count, number of interconnections (all gate inputs) and levels of logic (maximum number of gates through which any signal must pass) are compared below for all the circuits.

Circuit	No. of gates	No. of interconnections	Levels of logic
[1]	9	20	3
[2]	7	12	3
[3]	7	11	4

The best circuit would seem to be [2] or [3] depending on whether inter-connections or gate delays would be the more important criterion. Let us also look at the number of IC packages required to implement each circuit, since in practice we must count these for board layout. The following information is useful:

Type of gate	IC package designator	Gates per IC package
Inverter	7404	6 (hex)
2-input NAND	7400	4 (quad)
3-input NAND	7410	3 (triple)
4-input NAND	7420	2 (dual)

Circuit [2] requires one 7404 (leaving 3 spare NOT gates), one 7400 (leaving 1 spare NAND gate) and one 7410 (2 spare gates) – a total of 3 IC packages. The third circuit requires one 7404 (3 spare gates) and one 7400 (none spare), giving only 2 packages in total. So in terms of package count, circuit [3] is preferable to [2].

However, note that [1] requires one 7404 (2 spare gates), one 7420 (none spare) and one 7410 (none spare, using one of these as a 2-input NAND by connecting the unused input to HI) — the same number of packages as circuit [2]. Although [1] has a larger gate count it needs only the same number of ICs to implement it.

Note also that spare, that is unused, gates in packages may be an important factor to consider if they can be incorporated into another circuit which will share the circuit board.

The Quine–McCluskey minimisation method

An alternative to reducing the size of expressions using K-maps is the *tabular* minimisation method due to W. V. Quine and E. McCluskey. This is useful for expressions containing large numbers of variables, for which K-maps are impracticable. The method is systematic and is suitable for programming on a computer to give an automatic minimisation tool.

The aim of the tabular method is the same as that for K-maps, namely to extract the prime implicants from the Boolean expression and then to select the essential PIs, and sufficient of the remaining PIs, to cover all the terms in the expression.

Let us demonstrate the method by means of an example. The four-variable expression in the notes on circuit minimisation will be used, for two reasons — first, the techniques can be illustrated more easily while using a fairly small set of tables and second, a comparison can easily be made with the K-map method.

Example Tabular minimisation of the function:

$$F = \overline{A}\overline{B}\overline{C} + \overline{A}B + \overline{A}B\overline{C}\overline{D} + AC\overline{D}$$

(1) The first step is to convert the expression into canonical form, by expanding as follows

$$F = \overline{A}\overline{B}\overline{C}(D+\overline{D}) + \overline{A}B(C+\overline{C})(D+\overline{D}) + \ldots\ldots\ldots$$
$$= \overline{A}\overline{B}\overline{C}D + \overline{A}\overline{B}\overline{C}\overline{D} + \ldots\ldots\ldots\ldots (\text{expanded})$$

(2) The next step is to list the minterms (canonical product terms) in a table using 1s to represent uncomplemented and 0s to represent complemented variables, grouping the terms according to the number of 1s they contain. Thus $\overline{A}\overline{B}\overline{C}D$ would be written 0001 and grouped with the other terms

which contain only a single 1. Further, the terms are indexed by their decimal equivalent, so for example 0001 would be labelled '1' and 0101 ($\overline{A}B\overline{C}D$) labelled '5'.

The complete table is shown below:

	Decimal index	Binary equivalent of switching terms	
√	0	0000	group with no 1s
√	1	0001	
√	2	0010	group with one 1
√	4	0100	
√	5	0101	
√	6	0110	group with two 1s
√	10	1010	
√	7	0111	group with three 1s
√	14	1110	

(√ see explanation below).

(3) Now proceed to find, for each term in the table, those other terms which differ from that term in only one variable – this necessitates investigating the group below the current group in the table. So compare term 0 with all terms in the second group and compare, for example, term 5 with terms in the very last group.

The above process continues for each group from the first to the second last inclusive – since all terms in the last group will have been compared with all other terms during the process, this group need not be inspected again.

Whenever any pair of terms is found to differ in only one variable they are both ticked in the table. The pair is combined, replacing the variable which differs in the two terms by a dash (–), and inserting the combined pair into a new table. Thus, for example, term 1 and term 5 are combined to form 0–01, thus eliminating the variable B. The new table has the following entries (which should be checked by going through the first table systematically as described – note that all the terms in the first table have been ticked).

0,1	000-
0,2	00-0
0,4	0-00

1,5	0-01
2,6	0-10
2,10	-010
4,5	010-
4,6	01-0

5,7	01-1
6,7	011-
6,14	-110
10,14	1-10

(4) The second table is again organised into groups depending on the number of 1s in each term. The process carried out for table one is repeated for table two, where only terms with dashes in corresponding positions can be compared. Again, terms differing in only one variable are combined, the variable eliminated and an entry made in a new table. Terms which have been combined are ticked. In this example, all terms in table two will be ticked, and the new table is as follows.

0,1/4,5	0-0-
0,2/4,6	0--0
0,4/1,5	0-0-
0,4/2,6	0--0

2,6/10,14	--10
2,10/6,14	--10
4,5/6,7	01--
4,6/5,7	01--

(5) The same procedure is applied to this table, and so on until no more combinations can take place.

In the present example the third table yields no combinations so the process stops here. All remaining unticked entries in all the tables are the prime implicants of the expression.

Note that duplicate entries may arise, as in table three — in each case it is sufficient to include only one copy of each duplicated term. The third table therefore reduces to

0,1/4,5	or 0,4/1,5	0-0-	P
0,2/4,6	or 0,4/2,6	0--0	Q

2,6/10,14	or 2,10/6,14	--10	R
4,5/6,7	or 4,6/5,7	01--	S

The expression has four prime implicants, labelled P, Q, R and S. Compare this result with the Karnaugh map method for this example — in the tabular method all the prime implicants arise from the systematic procedure, whereas they are not always all obvious on K-maps.

(6) A prime implicant table is formed as shown below,

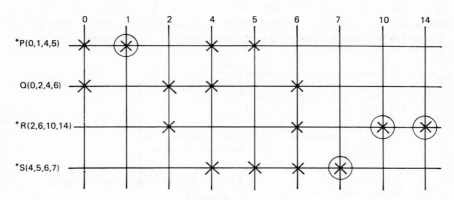

where each PI is given a name (in this example P, Q, R, and S) and a cross placed on the intersection of each PI row with any column representing a variable contained in that prime implicant.

To mark the essential prime implicants, look for columns containing exactly one cross. A circle is placed round the cross and the PI containing that variable has an asterisk placed alongside it. The PIs not asterisked are the other prime implicants. The essential prime implicants must be included in the answer, with a selection of the other PIs sufficient to cover all the terms in the expression.

In this example, since P, R and S are the essential prime implicants they are essential to the answer, but since they also cover all the terms in the original expression the fourth PI is not necessary, and can be excluded.

(7) The minimised expression is obtained by looking at the labelled entries for the finally selected PIs in the combination tables and writing down the

logical sum of the minimised terms, so in this example

$$F = \bar{A}B + \bar{A}\bar{C} + C\bar{D}$$

using the prime implicants P, R and S.

3.3 SEQUENTIAL LOGIC DESIGN

The Moore/Mealy model of sequential logic circuits in Fig. 3.11 shows that the essential difference between combinational and sequential circuits is the ability of the latter to remember previous outputs. Sequential circuits have a memory. As we shall see shortly, the important use of sequential circuits in computers is to implement registers; the simplest sequential circuits are basic memory elements.

The S–R flip-flop

The main components of a sequential circuit are a combinational logic part and a feedback path from output to input (via a memory element or delay path). The simplest sequential circuit is constructed with two NAND gates:

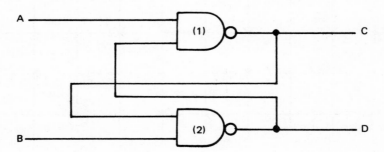

where the outputs are fed back and used as inputs. Let us analyse the operation of this simple circuit. We shall use the positive logic convention for all circuit analyses, in other words LO (low voltage) input is the same as 0, and a HI (high voltage) input is equivalent to 1. In analysing circuits it is important either to have a mental picture of the appropriate gate truth table (in this case for a 2-input NAND) or if this is found to be too difficult (at first) the truth table should be consulted (see Fig. 2.4). When A is 0, C must be 1. If B is 1, then both inputs to gate (2) are 1 and its output D is 0. If the input values are reversed, that is if A is 1 and B is 0, the outputs are similarly reversed: C and D are 0 and 1 respectively. If from either of these input cases (A = 1, B = 0 or A = 0, B = 1) B or A is changed so that both A and B are 1, then the values of C and D remain in their present states. Thus the circuit is capable of remembering which output, A or B, was last set to 0. The case A = B = 0 ensures that both C and D become 1.

The circuit is the basis of the *set-reset* (S–R) flip-flop which is used as a basic memory element. It is also called, in common with some other circuits which we shall be studying, a *bistable*: an element with two stable states. The S–R flip-flop circuit diagram is detailed in Fig. 3.20. It differs from the previous diagram in that the inputs are inverted (using single-input NAND gates).

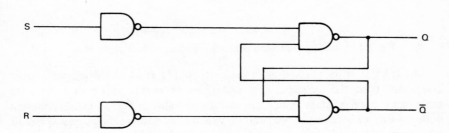

Fig. 3.20 – S–R flip-flop circuit diagram.

The normal state of the inputs is S = R = 0: in this state the outputs do not change but depend on which input, S or R, was previously applied. Applying an input means that it is changed to 1, then back to 0 again. In other words the input is *pulsed*: we shall look at the notion of pulses shortly. If S = 1 (and R = 0) then the output Q becomes 1. When S is returned to 0 the output Q remains at 1. If R = 1 (and S = 0) then Q changes to 0 and will remain so when R is returned to 0 (until the next time S is changed). In both cases (Q = 1 or Q = 0) the other output is the inverse of Q and is labelled $\bar{Q}$ for this reason. The operation of the S–R does not allow for the case when S = R = 1: this input combination must be avoided in practice. The behaviour of the S–R flip-flop is best summarised by using a state table or state diagram. Both are presented in Fig. 3.21. Note that Q_o labels the present internal state.

	Present state Q_o	Inputs		Outputs	
		S	R	Q	$\bar{Q}$
(a)	0	0	0	0	1
	0	0	1	0	1
	0	1	0	1	0
	0	1	1	X	X
*	1	0	0	1	0
	1	0	1	0	1
	1	1	0	1	0
*	1	1	1	X	X

(b)

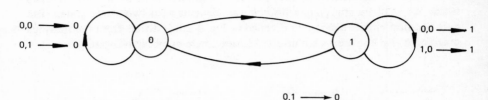

Fig. 3.21 – (a) State table and (b) state diagram for the S–R flip-flop.

In (a) the rows marked with an asterisk (*) label the disallowed input conditions (and the corresponding undefined outputs). Notice that the state table lacks the 'next state' column of the general table in Fig. 3.9(a). In common with the other bistables the next state is exactly the same as the output Q so this column would merely contain redundant information. More complex sequential circuits, with more than two internal states, would not in general have the same correspondence between their observable outputs and internal states. In (b) the order of inputs causing inter-state transitions is shown by $S = 1, R = 0 \rightarrow Q = 1$. The output $\bar{Q}$ is not specifically mentioned in this diagram, but is taken for granted. All flip-flops have complemented outputs available: they are often required in circuit implementations.

Pulses and clocks

The notions of pulses and clocks are very important in digital logic. Events in logic circuits take place over periods of time which we represent by means of *timing diagrams.* The basis of these diagrams is shown below:

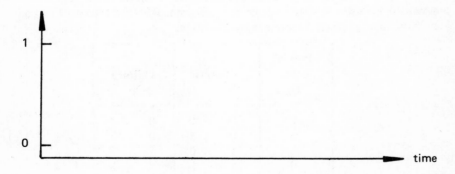

Time is represented along a horizontal axis, which is calibrated into suitably sized intervals, if necessary. Typically for logic circuits we would show time in a range between nanoseconds (10^{-9}s) and microseconds (10^{-6}s). The timing diagram is used to record the behaviour of voltage levels at a sampling point (an

input or output) in a circuit. Usually it suffices to represent only the two logic
levels, HI and LO (0 and 1 using positive logic) since, apart from very short
intervals, the sampling point will be at either one level or at the other. As a first
approximation, the *transition* between logic levels (0 to 1, or vice versa) may be
assumed to be instantaneous. As we shall see, there are circumstances in which it
is important to know the transition times in detail. For TTL logic, the levels 0
and 1 correspond approximately to 0 volts and 3.5 volts respectively (although
the voltage supply level V_{cc} is at 5 volts the output stage of TTL gates is such
that there is a voltage drop from V_{cc} through a resistor in the transistor ON
state — see Fig. 2.14).

Two consecutive logic level transitions constitute a *pulse,* either a *positive*
pulse as shown in Fig. 3.22(a), or a *negative* pulse (Fig. 3.22(b)). We shall deal
exclusively with positive pulses in our examples.

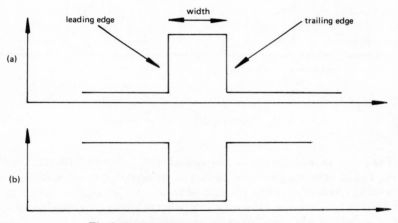

Fig. 3.22 – Pulses, (a) positive, (b) negative.

A $0 \to 1$ (LO $\to$ HI) transition is referred to as a *leading edge* (or *rising edge*
for positive pulses) and the $1 \to 0$ (HI $\to$ LO) transition as a *trailing edge* (or
falling edge). An important characteristic of a pulse is its *width* (or *duration*).

Some circuits are sensitive to a leading or trailing edge — these *edge-triggered*
devices will be described in the section on other flip-flops. Other flip-flops are
pulse-triggered which means that a whole pulse — two transitions — must be
applied to effect output changes. The S–R flip-flop, however, is perhaps best
described as *level-sensitive*: a HI logic level at S or R causes changes in Q. As
explained above, the mode of use of the S–R is to return the S or R level to LO
after the appropriate input has been asserted. One of the main operating problems
with logic circuits is *noise*. This manifests itself in the form of pulses, often
called *spikes* (or *glitches*), which have very small width. Such narrow pulses may

not be fast enough to cause changes and subsequent problems in pulse-triggered or level-sensitive circuits but they can produce unwanted output changes in edge-triggered devices. Elimination of noise is usually achieved by sound circuit construction techniques which include shielding from sources of external noise, good earthing contacts and the use of decoupling capacitors. Noise is a topic which will not be further discussed in this book.

In computer systems most (sequential) circuits are synchronised together by a common source of timing signals called a *clock*. The clock emits a stream of pulses at regular *intervals* as illustrated in Fig. 3.23.

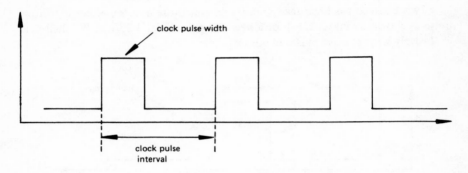

Fig. 3.23 — Pulses emitted by a clock source.

The pulses initiate changes in the various circuits to which the clock is connected, and in between pulses the circuits are allowed to operate or settle down. The settling down time can be the sum of many gate propagation delays, as for example in a basic parallel binary adder where carry bits are generated for each stage of the addition in turn. The *clock frequency*, which is the number of pulses emitted per unit of time, is chosen to suit the operating speed of all the circuits for which it provides timing signals: the speed of the slowest of these circuits ultimately determines the upper frequency limit of the clock source. Typical computer clock frequencies are of the order of 1 MHz (1 Megahertz = 10^6 pulses per second), in other words the clock pulse interval is about a microsecond or hundreds of nanoseconds. Pulse widths are typically about half the pulse interval, but may sometimes be much smaller than half. Frequency is measured only in terms of pulse intervals and says nothing about the width of clock pulses. Data sheets will, however, refer to the *duty cycle* which means the ratio of pulse width to interval. So a 50% duty cycle means that the width is half the pulse interval. The other flip-flops, which will be described next, are all provided with a *clock input* CK which is intended to be connected to a clock source. Such *clocked* flip-flops are the basis of the sequential circuits in modern computers.

Other types of flip-flop

There are two other types of flip-flop: the D and the J-K. Both are developments of the circuit described earlier for the S-R flip-flop. Their names, as for the S-R, are the labels used to denote the inputs. In the case of the D-type the name stands for *delay* (but is sometimes called *data*). However, the name J-K does not represent any words describing the J-K's mode of operation: the origin of J and K is unclear but they appear to have been chosen simply because they differ from S and R.

The circuit symbol for a flip-flop is a rectangular shape as explained in Section 2.1. Each type is distinguished by the input labels as illustrated in Fig. 3.24. This shows the approved circuit symbols for all three types, with some variations which are found in available IC packages.

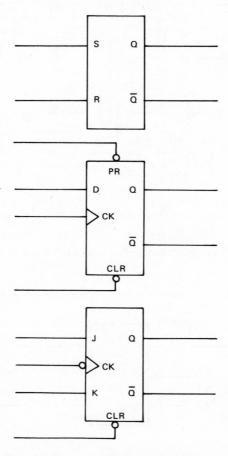

Fig. 3.24 – Circuit symbols for the main types of flip-flop.

In the diagram the S–R is unclocked, but the D and J–K both have clock inputs CK (the shape $>$ is also used to indicate the clock input). The D is provided with preset (PR) and clear (CLR) inputs which are active-LO (normal, inactive state HI) and asynchronous with the clock input: in other words if PR is set LO at any time the value of Q changes to 1, if CLR becomes LO the flip-flop state Q becomes 0. The CK input of the J–K is shown as active-LO: this will be explained shortly. Flip-flops are available in IC packages, usually in a dual configuration. The TTL IC version of the J–K in Fig. 3.24 is illustrated in Fig. 3.25. Its package designator is 7473. Notice the unconventional pin positions for V_{cc} and GND — usually they are assigned pins 14 and 7.

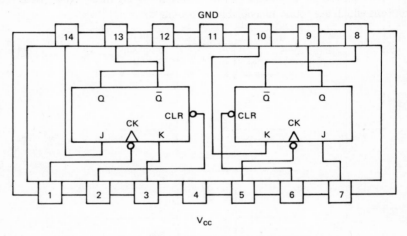

Fig. 3.25 — Dual J–K flip-flop package (7473).

D-type input is a single input which replaces the S and R inputs of Fig. 3.20, as illustrated below:

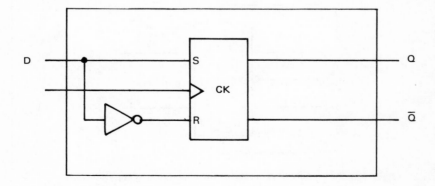

This arrangement is more accurately called a *gated latch* (or just a latch) because of its operation. Whenever the clock input is HI the output Q changes to the value currently at input D: Q follows D during a clock pulse. If the clock input is LO, Q cannot change. Effectively the clock is an *enable* input to the circuit. The latch operation is shown by the following timing diagrams:

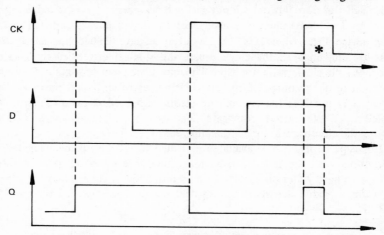

An important point to note is that if D changes while CK is HI, as during the clock pulse marked * , Q will change to follow it.

There are applications in which it is required to sample a data input at a particular time and to allow no further changes at the input to affect the output. A suitable way of achieving this is to sample on the edge of a pulse, during the very small time interval (typically a few nanoseconds) it takes the level to change from LO to HI or vice versa. The true D-type flip-flop is an edge-triggered

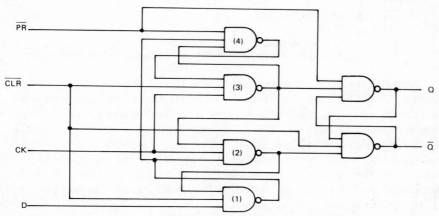

Fig. 3.26 – Circuit diagram of positive-edge triggered D-type flip-flop.

device rather than a latch, most commonly *positive-edge-triggered* meaning that it is the LO–HI transition of the clock input which causes changes. The circuit diagram of such a flip-flop is presented in Fig. 3.26. Clear and preset inputs, consistent with the D-type shown in Fig. 3.24, are included.

The circuit has the basic two NAND gates with crossed feedback paths as for the S–R flip-flop, but extra input gating is necessary to produce edge-triggered operation. Two cases must be considered in order to understand the operation of the device. First, when D is LO, the output of gate (2) becomes LO at the top of the leading edge of the clock pulse, and this effectively disables gate (1) throughout the clock pulse duration. Thus no subsequent changes in D can affect the circuit until the output from gate (2) is reset to HI by CK returning to LO. Second, if D is HI at the top of the leading clock edge, gates (2) and (4) are disabled by a LO output from gate (3), again preventing changes in D from affecting the circuit until the next leading edge.

This type of flip-flop is available in a dual form in an IC package designated 7474. Along with the pin configurations, data sheets describe the operation of flip-flops using a *function table* which is more like a truth table than the state table form introduced earlier. Such a description is exemplified for the 7474 in Fig. 3.27.

Inputs				Outputs	
Preset	Clear	Clock	D	Q	$\bar{Q}$
LO	HI	X	X	HI	LO
HI	LO	X	X	LO	HI
LO	LO	X	X	HI*	HI*
HI	HI	↑	HI	HI	LO
HI	HI	↑	LO	LO	HI
HI	HI	LO	X	Q_o	$\bar{Q}_o$

*nonstable configuration.

Fig. 3.27 – Function table for 7474 positive-edge-triggered D-type flip-flop.

As usual for manufacturers' data sheets, the information is presented in terms of LO and HI rather than 0 and 1. Don't care Xs are used to keep the table to a compact size, for example the first row in the table where preset is LO (and clear is HI) could be expanded to show each of the combinations of clock and D, but the essential information is that it doesn't matter what the values of clock and D are, so they are better shown as Xs. The basis of the circuit operation is denoted by the upward arrows, which mean that on the leading edge of the clock input, Q follows D. The state table form can be avoided by using Q_o and $\bar{Q}_o$ to denote previous states of Q and $\bar{Q}$.

The J–K flip-flop is a *master-slave* device. Each J–K contains effectively two

S–R flip-flops, as shown in Fig. 3.28. The idea of the master-slave operation is that either the master or the slave inputs are enabled at any time, but not both: the inverted clock input to the slave ensures this. Thus the J–K may be used as a one-bit storage element in a CPU register where certain operations require the register to supply a value while, within the same clock pulse interval, receiving a result from another source. Examples of such operations occur in a shift register, where on each clock pulse data is shifted one place right or left, and in an ALU where a register may be used to provide operands and receive results (a so-called *accumulator* register). The operation of the J–K of Fig. 3.25 (the 7473) is summarised by the function table in Fig. 3.29.

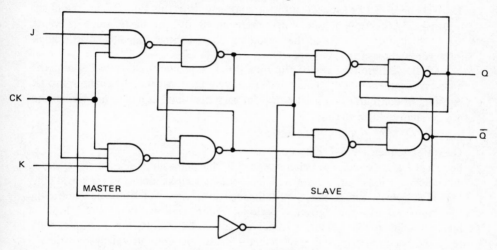

Fig. 3.28 – Circuit diagram of a master-slave J–K flip-flop.

	Inputs			Outputs	
Clear	Clock	J	K	Q	$\bar{Q}$
LO	X	X	X	LO	HI
HI	_⌐⌐_	LO	LO	Q_o	$\bar{Q}_o$
HI	_⌐⌐_	HI	LO	HI	LO
HI	_⌐⌐_	LO	HI	LO	HI
HI	_⌐⌐_	HI	HI	TOGGLE	

Fig. 3.29 – Function table for 7473 J–K flip-flop.

Apart from the asynchronous clear input, there are four input cases to consider. When J = K = LO the flip-flop stays in its former state (that is Q_o and $\bar{Q}_o$) after a clock pulse has been applied (signified by _⌐⌐_ in the table). J = HI,

K = LO sets the state of Q to HI; J = LO, K = HI resets Q to LO. Thus if J and K are both LO, or are different, they behave like S and R inputs, with the distinction that any changes in J and K while the clock input is HI do not become evident at Q until the trailing edge of the pulse: this is why the CK inputs in Fig. 3.25 have an active-LO appearance. The final input combination, J = K = HI, causes the flip-flop to TOGGLE, meaning that Q_o (and $\bar{Q}_o$) are complemented when a clock pulse is applied. Toggling is achieved by the extra feedback loops from Q and $\bar{Q}$ to the master input stage as in Fig. 3.28. The master-slave operation is perhaps best considered from the point of view of the clock pulse: while the clock is HI the master inputs are enabled but the slave inputs disabled. Thus changes can be clocked into the master during this time. When the clock returns to LO changes may no longer be recorded into the master since its inputs are disabled, but the state of the master flip-flop at the trailing edge of the clock pulse is now passed to the output Q.

Of the different types of flip-flop the J-K is the most versatile, and the most frequently used. It may be employed in constructing general-purpose registers in computers — one flip-flop for each bit — and as a memory element in any sequential logic circuit.

Designing sequential logic circuits

Sequential logic circuits are characterised by having a number of internal states by which previous outputs are remembered. As well as memory, they also contain logic which guides the circuit through a sequence of states depending on the input values applied over a period of time, as illustrated by the Moore/Mealy model in Fig. 3.11. Examples of sequential circuits are counters, shift registers, accumulators and (not directly related to computers) pattern detectors and traffic light sequencers.

The internal state of a sequential circuit is physically represented by as many flip-flops (one-bit memory elements) as are necessary. Each flip-flop corresponds at the abstract level to a *state variable*. To explain the distinction between internal states and state variables we may proceed as follows. Say a sequential circuit has 16 possible internal states. Each of these is represented on a state diagram by a single node (or circle). To implement the circuit, however, only 4 state variables (that is, flip-flops) are required since 16 different combinations of stored 1s and 0s can thus be represented. It is equally possible, without redundancy, to use as many as 16 flip-flops, each one corresponding to a state variable. Thus the n internal states of a sequential circuit can be realised by at least m flip-flops where m is the smallest integer greater than or equal to $\log_2 n$. If the minimum number (m) of flip-flops is used, the combinational logic required to *decode* the states may be substantially more complex to design than the logic necessary if n are used instead. Design complexity, however, is less important than implementation cost, so in practice the circuit using fewest IC packages — gates and flip-flops together — would probably be preferred.

There are two classes of sequential logic circuits, synchronous and asynchronous. The latter are composed of unclocked or free-running elements and are not our prime concern here. They are also more difficult to design than sequential circuits and require the use of *flow tables* and *excitation matrices* to represent their internal states and output transitions. An example of this class of circuit, and discussion of some of the associated problems, is given later. Synchronous circuits are our main concern — they predominate in computers. We shall see how they can be designed and some examples will be described.

Synchronous circuits consist of a number of flip-flops all connected to the same clock source and appropriate combinational logic interconnecting the flip-flops.

In general terms, the steps in sequential logic design are:

(1) Consider the given problem carefully and determine the number of internal states, inputs and outputs. A state diagram may be very useful for representing the desired operation of the circuit.

(2) Decide how many state variables, and therefore flip-flops, will be used in the implementation. Represent the circuit operation by means of a state table and for each state variable derive the associated combinational logic.

(3) Combine the separate flip-flops and their associated logic into the final circuit diagram, converting as appropriate for the available circuit components.

Typically the circuit implementation would be in terms of J–K flip-flops and NAND gates. To enable part (2) of the design process to be applied we begin by looking again at the state table for the S–R flip-flop in Fig. 3.21(a). It is possible to plot the information in the state table on a K-map, using the present state Q_0 as an input:

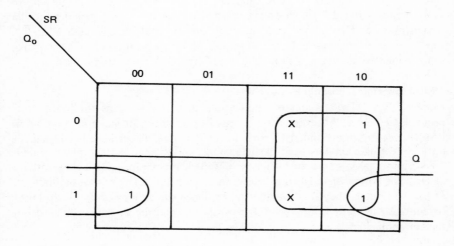

We obtain a minimised expression for the next state

$$Q = S + Q_0 \bar{R} \qquad (3.4)$$

called the *characteristic equation* for the S-R flip-flop.

A similar K-map can be drawn for the J-K flip-flop:

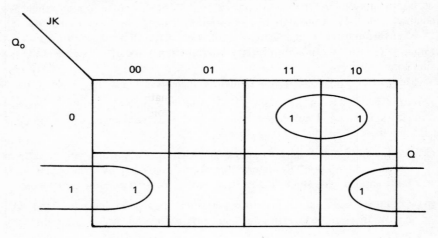

The characteristic equation for the J-K is

$$Q = \bar{Q}_0 J + Q_0 \bar{K} \qquad (3.5)$$

Equations (3.4) and (3.5) may be used to help implement sequential circuits in terms of (clocked) S-R or J-K flip-flops respectively by enabling the appropriate combinational logic to be extracted from the state table produced in the design process. They allow us to identify the logic which acts as S-R or J-K inputs for each state variable. An example will illustrate the design of synchronous sequential circuits and show how the characteristic equation is used.

Design example – modulo-8 counter

Counters have a wide range of applications in digital systems for counting or sequencing events. An event is represented by a pulse, for example a radiation counter records the number of particles of radiation which have been incident on a collecting chamber: each particle causes an avalanche effect which is converted into a pulse of electrical energy. Counters are employed in digital watches and clocks, where the triggering pulse is derived from quartz crystal oscillations, and in computers they may be used to sequence the actions of the control unit or control the steps in a serial shift operation according to timing pulses from the common clock.

They are useful examples for illustrating the design and operation of sequential circuits because they are simple to explain. Counters also demonstrate well the sequencing action of sequential circuits. There are various types — binary up and down counters and ring counters being particularly important — but they may be devised to count in any code we like (for example, Gray code). Our design example is a (synchronous) re-cycling modulo-8 binary up counter, meaning that it counts in an increasing sequence through 8 states then repeats the sequence. It is usual to count from 0, so the 8 states correspond (in binary) to the decimal values 0 through 7. As our first step in the design process we must consider how many internal states, inputs and outputs the circuit will have. There is only one input — the source of pulses which are being counted. The outputs must show the current count value, which is the same as the internal state. In the next section we shall design a (combinational) decoder circuit which can translate the internal state for representation on a seven-segment light-emitting diode (LED) display. There will be eight internal states, usefully represented on the state diagram of Fig. 3.30.

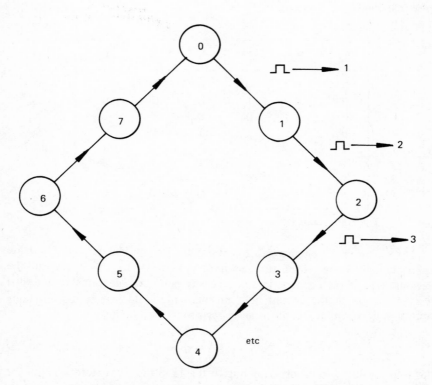

Fig. 3.30 — State diagram of re-cycling modulo-8 up counter.

This shows the eight states as eight nodes, and the cyclic nature of this counter. An input pulse causes the counter to sequence to its next state and to output the new states.

How many state variables, that is flip-flops, do we require to implement this counter? A minimum of three is required since there are eight states to represent. With three state variables we can use a pure binary representation of the eight states. Let us proceed with the design using three state variables, which we shall call A, B and C. The logic circuit will consist of three flip-flops and appropriate combinational logic. To find out what logic is required could be done intuitively, given a good understanding of the operation of the flip-flops to be used in the design. However, a methodical approach is possible making use of the characteristic equation of the flip-flops we intend to use. We begin by writing down a state table in terms of the state variables A, B and C. This is shown in Fig. 3.31. The input pulse is not represented; it is assumed for all synchronous circuits that changes can occur only when triggered by a pulse at the clock input of the flip-flops, so it is not necessary to show this input in the table. A_0, B_0 and C_0 represent the previous states of A, B and C before a pulse is applied.

A_0	B_0	C_0	A	B	C
0	0	0	0	0	1
0	0	1	0	1	0
0	1	0	0	1	1
0	1	1	1	0	0
1	0	0	1	0	1
1	0	1	1	1	0
1	1	0	1	1	1
1	1	1	0	0	0

Fig. 3.31 — State table for re-cycling modulo-8 binary up counter.

Let us design the circuit with J–K flip-flops. We consider each state variable separately and write down the relationship between its new state and the previous state of the counter. In the same way as for combinational logic design we look for 1s in the output column and write a Boolean sum-of-products expression of the input combinations giving rise to the 1s. Thus:

$$A = \bar{A}_0 B_0 C_0 + A_0 \bar{B}_0 \bar{C}_0 + A_0 \bar{B}_0 C_0 + A_0 B_0 \bar{C}_0 \qquad (3.6)$$

(the four product terms corresponding to the 1s in the 'next state' column for A).

The characteristic equation of the J–K flip-flop is (substituting A for Q in equation (3.5)):

$$A = \bar{A}_o J_A + A_o \bar{K}_A$$

Re-writing (3.6),

$$A = \bar{A}_o (B_o C_o) + A_o (\bar{B}_o \bar{C}_o + \bar{B}_o C_o + B_o \bar{C}_o) \tag{3.7}$$

Equation (3.7) tells us that the required logic at the J input of the flip-flop representing A is $B_o C_o$, and the logic at the K input is the inverse of $(\bar{B}_o \bar{C}_o + \bar{B}_o C_o + B_o \bar{C}_o)$. Evaluating the inverse using Boolean algebra, or more easily by means of a K-map (the inverse of a plotted function consists of the blank squares), we obtain $B_o C_o$ as the K input logic. That is,

$$J_A = K_A = B_o C_o$$

A more direct way of extracting the J and K input logic is to inspect the state table for $0 \rightarrow 1$ and $1 \rightarrow 0$ transitions of the state variables. Input combinations giving rise to $0 \rightarrow 1$ transitions correspond to $\bar{Q}_o J$ whilst those producing $1 \rightarrow 0$ changes correspond to $Q_o K$ (K, not $\bar{K}$: this will give the K logic directly). Using this technique for A in our example,

$$0 \rightarrow 1 \text{ transitions: } \quad \bar{A}_o J_A = \bar{A}_o B_o C_o$$
$$J_A = B_o C_o$$
$$1 \rightarrow 0 \text{ transitions: } \quad A_o K_A = A_o B_o C_o$$
$$K_A = B_o C_o$$

giving J_A and K_A as before.

The logic for state variable A in the modulo-8 counter is:

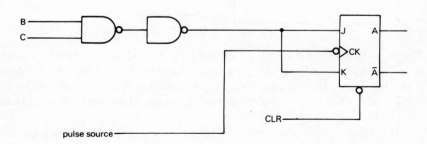

where B and C are derived directly from the Q outputs of flip-flops B and C.

Repeating the procedure, this time for state variable B:

$$0 \rightarrow 1: \bar{B}_o J_B = \bar{A}_o \bar{B}_o C_o + A_o \bar{B}_o C_o$$
$$J_B = \bar{A}_o C_o + A_o C_o = C_o$$
$$1 \rightarrow 0: B_o K_B = \bar{A}_o B_o C_o + A_o B_o C_o$$
$$K_B = \bar{A}_o C_o + A_o C_o = C_o$$

The circuit for B is:

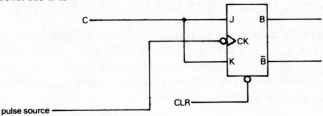

For state variable C:

$$0 \rightarrow 1: \bar{C}_o J_c = \bar{A}_o \bar{B}_o \bar{C}_o + \bar{A}_o B_o \bar{C}_o + A_o \bar{B}_o \bar{C}_o + A_o B_o \bar{C}_o$$
$$= \bar{C}_o$$
$$J_c = 1 \text{ (or HI)}$$
$$1 \rightarrow 0: C_o K_c = \bar{A}_o \bar{B}_o C_o + \bar{A}_o B_o C_o + A_o \bar{B}_o C_o + A_o B_o C_o$$
$$= C_o$$
$$K_c = 1 \text{ (or HI)}$$

The circuit for C is:

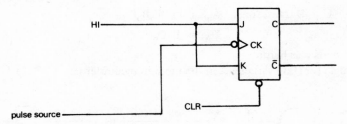

The complete counter circuit is obtained by combining the separate circuits, as in Fig. 3.32. Strictly speaking, our state diagram of Fig. 3.30 should indicate a starting state. The starting state of the implementation we have derived is ABC = 000, which is obtained by activating the asynchronous CLR input to the circuit.

The same design technique may be used for any synchronous sequential logic circuit. Variations on the counter such as different modulo number, down counting instead of up, or self-stopping instead of re-cycling can be taken account of in the state diagram and state table – and the design follows as before. Further design examples are given in the next chapter.

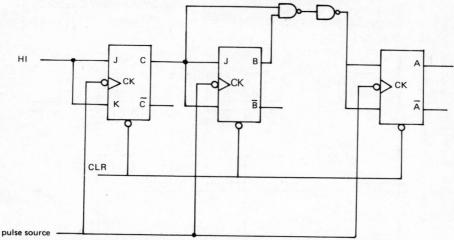

Fig. 3.32 – Circuit diagram of synchronous modulo-8 binary counter.

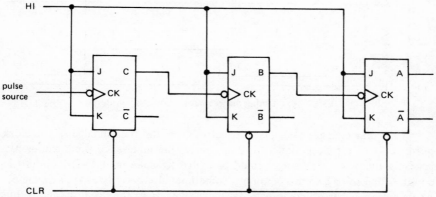

Fig. 3.33 – Circuit diagram of an asynchronous modulo-8 binary counter.

An asynchronous counter

As an example of an asynchronous sequential circuit, a modulo-8 binary counter is illustrated in Fig. 3.33. The circuit consists of three J–K flip-flops as for the synchronous version, but the pulse source is connected only to one flip-flop: the flip-flops in the circuit are not synchronised by a common signal. This counter operates according to the state diagram of Fig. 3.30 – that is it counts 0 through 7 and re-cycles – but the state diagram does not indicate the internal changes in the circuit following an input pulse. Since there is no synchronisation, changes are propagated through the circuit rather than affecting all the flip-flops simultaneously. For this reason, the asynchronous counter is often referred to as a *ripple-through* counter.

Each J–K operates in toggle mode, that is whenever a pulse is received at CK the output is complemented. The effect is summarised by the timing diagrams in Fig. 3.34.

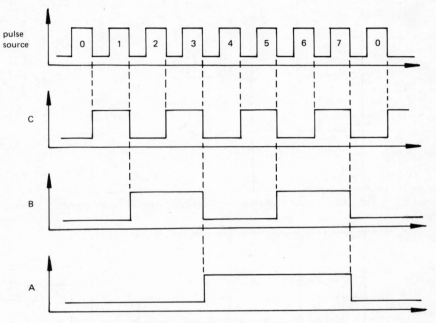

Fig. 3.34 – Timing diagrams for modulo-8 counter.

Note that changes take place effectively on the trailing edge of the clock pulse for J–K flip-flops. Each flip-flop divides the number of input pulses successively by two (extra stages could be added to make modulo-16, modulo-32 counters and so on), so this counter is sometimes alternatively called a *divide-by-two* counter. It has no combinational logic external to the flip-flops and is thus minimal in terms of IC packages. However, it has a disadvantage: it takes longer to settle down into a new state than the synchronous counter when a new pulse arrives. This is because flip-flops, like gates, have a *propagation delay* time. Typically, a TTL flip-flop takes 30 nanoseconds before its output has changed in response to the trailing edge of a pulse at its CK input. This is illustrated below:

Such a delay is of course present in synchronous counters, but the total delay is exactly the same: 30 nanoseconds. For the asynchronous counter, however, consider the pulse labelled 7 in Fig. 3.34. There will be three consecutive 30 ns delays before the counter changes from 111 to the 000 state, a total of 90 ns.

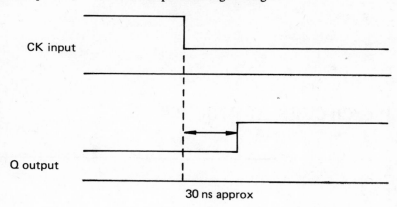

30 ns approx

For larger asynchronous counters even longer delays occur. When using an asynchronous counter, these longer settling times must be accounted for when sampling the state of the counter, otherwise erroneous readings could be taken. A problem with logic circuits in general is the possibility of erroneous operation because of circuit *hazards* — faults which arise because of the interaction of changing internal signals in circuits. These are particularly likely to occur in asynchronous circuits and can be difficult to eliminate. Hazards in combinational circuits will be discussed briefly in the next chapter.

Further remarks on sequential design

Synchronous sequential circuits, for which a design method was described, are in practice more commonly used. While the design method outlined is sufficient for the majority of requirements, two refinement techniques should be mentioned.

Just as combinational circuits can be minimised, so techniques can be applied to reduce both the number of flip-flops and the amount of combinational logic in sequential circuits. In more complex design problems two or more internal states may be equivalent, that is given the same input sequences they produce exactly the same outputs (even though their next internal states are not the same). *Internal state reduction* (or *state minimisation*) can produce a design solution with fewer flip-flops, though this may have the effect of increasing the associated combinational logic.

The technique of *state assignment* is used to assign binary codes optimally to the internal states identified in the design process. This can have the effect of reducing the combinational logic in the final circuit. For simple examples, however, it is possible to rely on intuitive methods.

CHAPTER 4

Logic circuits in practice

4.1 DESIGN EXAMPLES

We begin this chapter by presenting further design examples of both combinational and sequential logic circuits, three of each type. They will consolidate the design techniques introduced in Chapter 3 as well as providing examples of logic circuits which may be put to practical use. The present chapter continues with a discussion of some problems which are encountered in using logic components, followed by some remarks on analysing (rather than synthesising) circuits, and lastly there comes a section dealing with some other approaches to logic circuit implementation.

Priority encoder

A priority encoder has several input lines which are normally at 0. When any of the inputs goes to 1 the encoder outputs a value identifying that input line. Furthermore, the inputs have each a different priority, and if two of them are at 1 the encoder outputs the identity of the higher priority line. The encoder has another output line which indicates whether any input line is at 1. Such an encoder could be used in a priority interrupt scheme whereby peripheral devices are connected to a CPU and require to interrupt its operation on completion of a data transfer. The highest priority device must be given attention if more than one device signals completion at the same time.

The circuit for a 4-input priority encoder is to be designed. In order to identify uniquely one of 4 lines, 2 outputs are required at least. These we shall call the address lines for the inputs. A black-box diagram of the encoder is shown below:

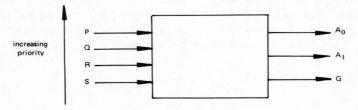

P, Q, R and S are the input lines and have relative priorities as indicated in the diagram, A_0 and A_1 the address lines, and G the output which tells if any input line is set to 1. To design the required combinational logic the following truth table is constructed:

P	Q	R	S	G	A_1	A_0
0	0	0	0	0	X	X
0	0	0	1	1	0	0
0	0	1	X	1	0	1
0	1	X	X	1	1	0
1	X	X	X	1	1	1

The use of don't care conditions in both input and output columns makes the table much more compact and readable than if they are expanded to show all combinations explicitly.

Treating each output separately we draw the following K-maps:

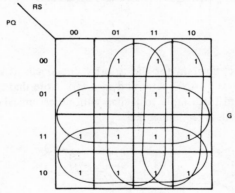

which gives simply

$$G = P + Q + R + S$$

so $A_1 = P + Q$

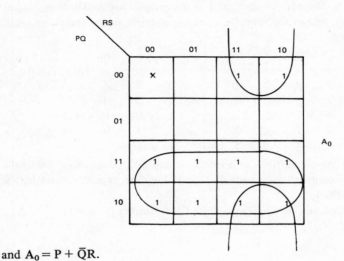

and $A_0 = P + \bar{Q}R$.

    Thus the circuit diagram for the priority encoder is as shown in Fig. 4.1. It is drawn in terms of NAND gates and inverters. Note that above it is possible to plot directly from truth table to K-map without canonical expansion (with a little practice) and this is recommended.

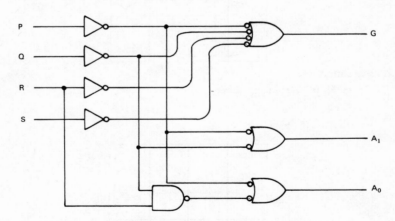

Fig. 4.1 – Circuit diagram of 4-line to 2-line priority encoder.

(An 8-line to 3-line priority encoder, for example, is available as a TTL MSI package, designated 74148).

Parity generator/checker

Parity generators and checkers are used in the input/output part of a computer system. It is usual for information to be transmitted in groups of bits representing a character or symbol. Errors may occur in transmission; therefore it is important to be able to detect their occurrence at the receiving end, and better still to correct the error as well. Single-bit errors may be detected by a parity bit sent with the data. There are two basic schemes: if the number of 1s in the data is even, a parity bit of 0 is included, otherwise parity sent is 1 (this is called *even parity*) or the other way round, namely parity is generated to ensure the total number of 1s sent is odd (*odd parity*). At the receiving end, parity is checked by looking for either an even or an odd number of 1s.

To illustrate the generation of even parity, consider a 4-bit data word, ABCD. A single parity bit P is to be produced as shown by the black-box diagram:

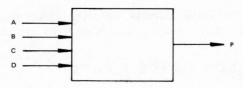

The truth table for the generator is as follows:

A	B	C	D	P
0	0	0	0	0
0	0	0	1	1
0	0	1	0	1
0	0	1	1	0
0	1	0	0	1
0	1	0	1	0
0	1	1	0	0
0	1	1	1	1
1	0	0	0	1
1	0	0	1	0
1	0	1	0	0
1	0	1	1	1
1	1	0	0	0
1	1	0	1	1
1	1	1	0	1
1	1	1	1	0

Using a K-map directly:

AB \ CD	00	01	11	10
00		1		1
01	1		1	
11		1		1
10	1		1	

No reduction can be effected. However, this table has a form similar to the one on page 71 for the binary full-adder sum. Writing out the Boolean expression for P:

$$P = \bar{A}\,\bar{B}\,\bar{C}\,D + \bar{A}\,\bar{B}\,C\,\bar{D} + \bar{A}\,B\,\bar{C}\,\bar{D} + \ldots$$

which can be regrouped as

$$P = (\bar{A}\,\bar{B} + A\,B)(\bar{C}\,D + C\,\bar{D}) + (\bar{A}\,B + A\,\bar{B})(\bar{C}\,\bar{D} + C\,D)$$
$$= (\overline{A \oplus B})(C \oplus D) + (A \oplus B)(\overline{C \oplus D})$$
$$= X \oplus Y$$

where $\quad X = A \oplus B \quad$ and $\quad Y = C \oplus D$

Another way of writing this is simply

$$P = A \oplus B \oplus C \oplus D$$

(brackets may be placed anywhere within this expression).

The circuit is a cascade of EOR gates as in Fig. 4.2. Note that the generator circuit can easily be extended for any number of data bits. A realistic word

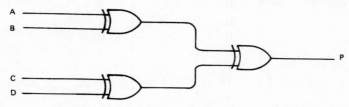

Fig. 4.2 – Circuit diagram of 4-bit parity generator.

length over which to generate priority is 8 bits (as for the ASCII international code to represent characters) including 1 bit for parity. In this case P would be given by

$$A \oplus B \oplus C \oplus D \oplus E \oplus F \oplus G$$

where A–G are the seven data bits.

Parity checking uses the same form of circuit, except that at the receiving end the parity bit is itself an input. A general-purpose 9-bit parity generator/checker is available in TTL MSI form, designated 74180, with 8 data bits and a ninth input (actually 2 bits) used for selecting odd or even parity if the circuit is used for generation and for the parity bit itself when the circuit is used as a checking circuit.

Modulo-n counters

In Section 3.3 a modulo-8 binary counter was designed. The eight states could be represented by three flip-flops leaving no spare states. In general, modulo-n counters require a minimum of $\log_2 n$ flip-flops rounded up to the nearest integer but unless n is a power of two there will be spare states left over. An example is a modulo-10 counter which needs 4 flip-flops. The state table is shown below:

A_o	B_o	C_o	D_o	A	B	C	D
0	0	0	0	0	0	0	1
0	0	0	1	0	0	1	0
0	0	1	0	0	0	1	1
0	0	1	1	0	1	0	0
0	1	0	0	0	1	0	1
0	1	0	1	0	1	1	0
0	1	1	0	0	1	1	1
0	1	1	1	1	0	0	0
1	0	0	0	1	0	0	1
1	0	0	1	0	0	0	0
1	0	1	0	X	X	X	X
1	0	1	1	X	X	X	X
1	1	0	0	X	X	X	X
1	1	0	1	X	X	X	X
1	1	1	0	X	X	X	X
1	1	1	1	X	X	X	X

Only ten states are required — the remaining 6 are spare and can be used as don't care conditions.

This is a cyclic counter as before — returning to state 0000 after 1001. This counter can be used in digital clocks or watches along with modulo-6 and modulo-3 counters (for a 24-hour clock) as illustrated below:

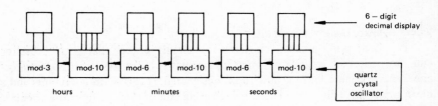

The oscillator provides a 1-second interval clock pulse to the least significant counter. Typically the crystal will oscillate at a frequency of 32768 Hz so the pulse rate will be divided down by a counter arrangement to provide the 1-second timing. Each counter triggers the next more significant one by a pulse when it recycles from its maximum count to zero. This simple arrangement gives a ripple-through effect for the digital clock. Extra logic must be included (though it is not shown) to cause the modulo-10 hours counter to recycle from 3 to zero (instead of from 3 to 4) when the most significant modulo-3 counter is already at 2. This can be taken account of in the design of the modulo-10 hours counter, by including an input derived from the mod-3 counter in the state table. A 6-digit decimal display would normally be provided. Suitable logic to drive the displays will be designed in the next example.

Meantime the design of the synchronous modulo-10 counter may proceed from the state table above. The don't care conditions are included to improve the solution where appropriate. For state variable A, we look for $0 \rightarrow 1$ and $1 \rightarrow 0$ transitions as before. We shall design the circuit using J-K flip-flops:

$$0 \rightarrow 1: \quad \bar{A}_o J_A = \bar{A}_o B_o C_o D_o$$

$$1 \rightarrow 0: \quad A_o K_A = A_o \bar{B}_o \bar{C}_o D_o \ (+ A_o \bar{B}_o C_o \bar{D}_o + A_o \bar{B}_o C_o D_o +$$

$$A_o B_o \bar{C}_o \bar{D}_o + A_o B_o \bar{C}_o D_o + A_o B_o C_o \bar{D}_o + A_o B_o C_o D_o)$$

(don't care conditions are in brackets).

In the case of the first equation there are no don't care conditions. Thus we may write

$$J_A = B_o C_o D_o$$

Eliminating A_o from the second equation, the solution for K_A is best found by plotting the right hand side on a 3-variable K-map:

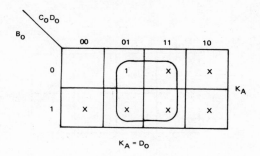

$$K_A = D_o$$

Similarly for state variables B, C and D we derive:

$$0 \rightarrow 1: \quad \bar{B}_o J_B = \bar{A}_o \bar{B}_o C_o D_o \; (+ A_o \bar{B}_o C_o \bar{D}_o + A_o \bar{B}_o C_o D_o)$$

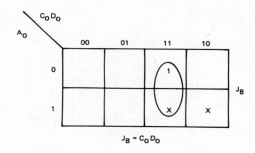

$$J_B = C_o D_o$$

$$1 \rightarrow 0: \quad B_o K_o = \bar{A}_o B_o C_o D_o \; (+ A_o B_o \bar{C}_o \bar{D}_o + A_o B_o \bar{C}_o D_o$$
$$+ A_o B_o C_o \bar{D}_o) + A_o B_o C_o D_o)$$

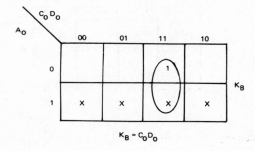

$$K_B = C_o D_o$$

$$0 \rightarrow 1: \ \overline{C}_o J_c = \overline{A}_o \overline{B}_o \overline{C}_o D_o \ + \ \overline{A}_o B_o \overline{C}_o D_o \ (+ \ A_o B_o \overline{C}_o \overline{D}_o \ +$$
$$A_o B_o \overline{C}_o D_o)$$

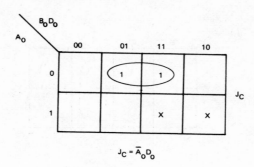

$$J_C = \overline{A}_o D_o$$

$$1 \rightarrow 0: \ C_o K_c = \overline{A}_o \overline{B}_o C_o D_o \ + \ \overline{A}_o B_o C_o D_o \ (A_o \overline{B}_o C_o \overline{D}_o \ +$$
$$A_o \overline{B}_o C_o D_o + A_o B_o C_o \overline{D}_o + A_o B_o C_o D_o)$$

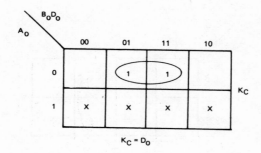

$$K_C = D_o$$

Lastly, the K-maps for D are both:

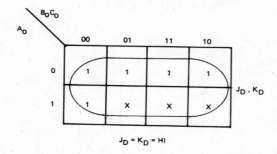

$$J_D = K_D = HI$$

The circuit diagram for the modulo-10 counter is shown in Fig. 4.3.

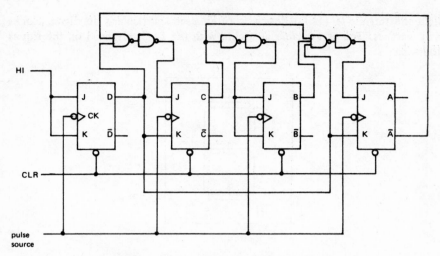

Fig. 4.3 – Circuit diagram of modulo-10 binary counter.

A simple method of providing the clock pulses for the next more significant counter in the 24-hour clock would be to NAND the $\bar{Q}$ outputs of the flip-flops together as shown below for the modulo-10 counter:

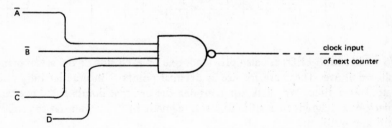

Whenever the counter is in the zero state the output of the NAND gate is LO; otherwise it is HI. When the counter changes from zero to one the leading edge of a pulse is produced. At the change from the maximum count to zero the trailing edge of the pulse triggers the next counter.

A TTL MSI modulo-10 counter is available in a positive-edge-triggered variety, designated 74160.

BCD to seven-segment LED decoder

Binary-coded-decimal (BCD) employs four bits to represent the decimal digits 0–9. An example of its use is in the digital clock described in the previous section. Each modulo-10 counter is capable of storing a BCD value (the other counters a subset of the possible values). For display purposes a seven-segment light-emitting-diode (LED) device is often used, particularly in digital clocks and watches. This is available as a chip with the LEDs arranged on the top as follows:

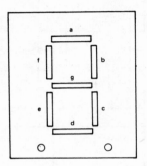

The seven segments a–g (the labels are for identification and do not appear on the chip) may be lit or unlit in combinations to form all the decimal digits, as:

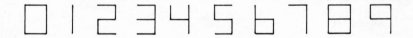

Two small round LEDs are also provided, one on each side of the seven segments as shown above. These are for use as decimal points – in practice only one can be used at a time. We shall not consider the decimal points for this example. A black-box diagram of a BCD to seven-segment LED decoder has the following inputs and outputs:

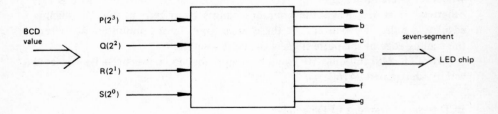

P, Q, R and S are the four bits of the BCD value, a–g the output lines which will cause the appropriate segments to be lit (considered active-HI in our example). The truth table for our decoder design follows:

decimal	P	Q	R	S	a	b	c	d	e	f	g
0	0	0	0	0	1	1	1	1	1	1	0
1	0	0	0	1	0	1	1	0	0	0	0
2	0	0	1	0	1	1	0	1	1	0	1
3	0	0	1	1	1	1	1	1	0	0	1
4	0	1	0	0	0	1	1	0	0	1	1
5	0	1	0	1	1	0	1	1	0	1	1
6	0	1	1	0	0	0	1	1	1	1	1
7	0	1	1	1	1	1	1	0	0	0	0
8	1	0	0	0	1	1	1	1	1	1	1
9	1	0	0	1	1	1	1	0	0	1	1
	1	0	1	0	X	X	X	X	X	X	X
	1	0	1	1	X	X	X	X	X	X	X
	1	1	0	0	X	X	X	X	X	X	X
	1	1	0	1	X	X	X	X	X	X	X
	1	1	1	0	X	X	X	X	X	X	X
	1	1	1	1	X	X	X	X	X	X	X

including don't care conditions for the six unused input combinations. The K-map and reduced expression for each output are listed below:

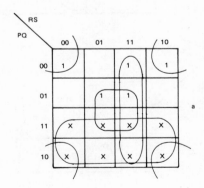

$$a = P + RS + \bar{Q}\,\bar{S} + Q\,S$$

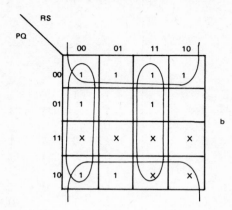

$$b = \bar{Q} + \bar{R}\bar{S} + RS$$

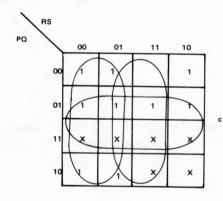

$$c = Q + \bar{R} + S$$

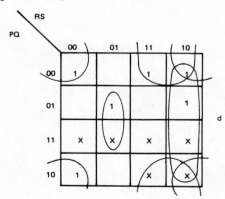

$$d = \bar{Q}\bar{S} + \bar{Q}R + R\bar{S} + Q\bar{R}S$$

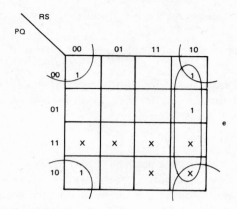

$$e = \bar{Q}\,\bar{S} + R\,\bar{S}$$

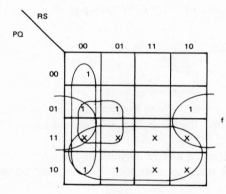

$$f = P + Q\,\bar{R} + \bar{R}\,\bar{S} + Q\,\bar{S}$$

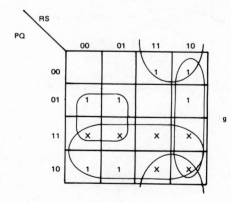

$$g = P + Q\,\bar{R} + R\,\bar{S} + \bar{Q}\,R$$

Note that the different shapes for 6 and 9, namely,

could be used in the design, producing slightly different expressions for a and d.

The logic circuit for the decoder is shown in Fig. 4.4. A similar TTL MSI component, the 7448, is available.

Pattern detector

A pattern detector is a finite-state machine (FSM) which inputs an indefinitely long string of symbols and outputs a special symbol whenever it recognises a particular pattern. Let us consider a simple version of the problem: we wish to be able to detect the binary pattern 0101 on an input line X, outputting a 1 on line R if the pattern is recognised (otherwise 0s will be output). The input bits are to be clocked into the pattern recogniser as illustrated below:

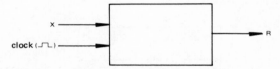

Since we shall design a synchronous sequential circuit to implement the recogniser, we can assume that the clock line is fed to the CK input of each flip-flop in the circuit.

Let us begin by drawing a state diagram to show how many internal states are required, and how the recogniser sequences between them: Four states are

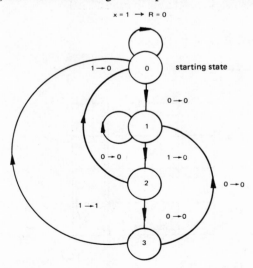

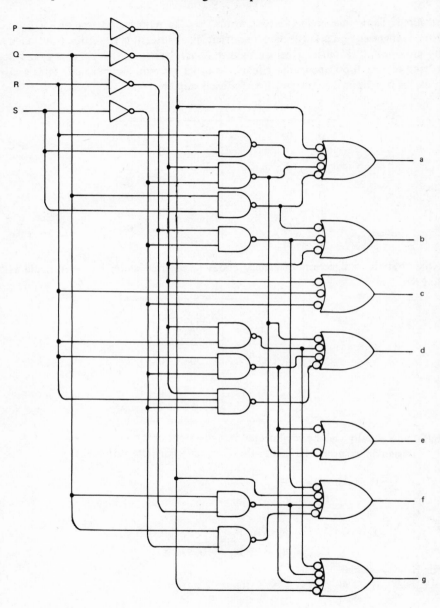

Fig. 4.4 – Circuit diagram of BCD to seven-segment LED decoder.

required. The sequence 0-1-2-3-0 means that the required pattern of 0101 has been recognised and a 1 will then be output. Each interstate transition is actioned by an input clock pulse. Thus we require 2 state variables (A and B, say) and 2 corresponding flip-flops in the circuit. As usual we write down a state table and look for $0 \rightarrow 1$ and $1 \rightarrow 0$ transitions for each variable:

X	A_o	B_o	A	B	R
0	0	0	0	1	0
1	0	0	0	0	0
0	0	1	0	1	0
1	0	1	1	0	0
0	1	0	1	1	0
1	1	0	0	0	0
0	1	1	0	1	0
1	1	1	0	0	1

Note that the assignment of binary codes to internal states has been made as follows:

State	Code A	B
0	0	0
1	0	1
2	1	0
3	1	1

but could equally well be re-allocated in a different order.

Designing in terms of J-K flip-flops, we have for state variable A:

$$0 \rightarrow 1: \bar{A}_o J_A = X \bar{A}_o B_o$$
$$J_A = X B_o$$

$$1 \rightarrow 0: A_o K_A = XA_o \bar{B}_o + \bar{X} A_o B_o + X A_o B_o$$
$$K_A = B_o + X\bar{B}_o = B_o + X$$

For B:

$$0 \rightarrow 1: \bar{B}_o J_B = \bar{X} \bar{A}_o \bar{B}_o + \bar{X} A_o \bar{B}_o$$
$$J_B = \bar{X}$$

$$1 \rightarrow 0: B_o K_B = X \bar{A}_o B_o + X A_o B_o$$
$$K_B = X$$

(without the need for K-maps)

Extra combinational logic is required for output R. From the state table

$$R = X A_o B_o$$

Thus the complete circuit can be drawn, as in Fig. 4.5 (CLR inputs are not shown)

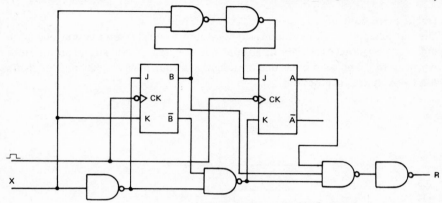

Fig. 4.5 − Circuit diagram of 0101 pattern recogniser.

The problem can be tackled in quite a different way, using an ad hoc technique. Notice that what we are really doing is interposing a logic circuit on the stream of input bits and looking at a window, four bits wide.

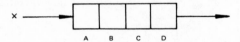

This is just a *right-shift register* arrangement, a group of flip-flops connected together such that at each clock pulse the bits all move one flip-flop to the right (left-shift registers can be built similarly). The J and K inputs of each flip-flop are derived from the Q and $\bar{Q}$ outputs respectively of the flip-flop on the left. To recognise 0101 we simply add NAND gating as shown in Fig. 4.6.

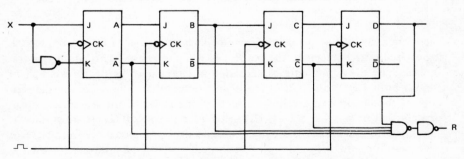

Fig. 4.6 − Alternative circuit diagram of 0101 pattern recogniser.

Using the shift register arrangement a programmable recogniser could be constructed. The pattern to be recognised could be set up in a 4-bit register and comparison logic designed so that R would output 1 only when that register and ABCD (as in Fig. 4.6) contained the same pattern.

Traffic lights controller

The last example in this section is slightly more complex than the previous ones, and requires more explanation, although it concerns a realistic problem from everyday life. The problem is one of designing a controller to sequence the lights at a two-way traffic junction.

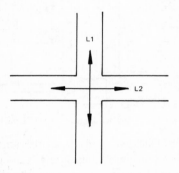

There are two distinct sets of lights, L1 and L2, each with a green, amber and red lamp. The lights work in a sequence which may be illustrated as follows:

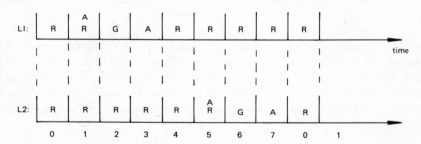

There are eight different combinations of lit lamps, labelled 0–7. A sequential logic circuit to implement the sequences will thus have eight internal states. In each state the appropriate combination of lamps must be lit, for example in state 0 both L1(R) and L2(R) are lit but none other. The lamps are the outputs of the circuit and are lit by *decoding* the states of the sequencer appropriately. Let us say that a 1 (or HI) lights a lamp, and a 0 (or LO) extinguishes it. The input to the circuit is a pulse source to drive the sequencer from state to state, but it has this distinction: it must have irregular pulse intervals because the eight states will not all have equal durations. In particular state 0 will be short whereas state 2 will be much longer. The problem of this timing pulse will be dealt with shortly.

Meantime, assuming a suitable pulse source we now proceed to design the sequencer and its lamp decoding logic. Eight internal states can be suitably provided by the synchronous modulo-8 counter which was descibed in Section 3.3. We shall not reproduce the state table nor the design here. An amalgamated truth table for the lamp logic is, however, necessary:

State	A	B	C	L1 R	L1 A	L1 G	L2 R	L2 A	L2 G
0	0	0	0	1	0	0	1	0	0
1	0	0	1	1	1	0	1	0	0
2	0	1	0	0	0	1	1	0	0
3	0	1	1	0	1	0	1	0	0
4	1	0	0	1	0	0	1	0	0
5	1	0	1	1	0	0	1	1	0
6	1	1	0	1	0	0	0	0	1
7	1	1	1	1	0	0	0	1	0

Using K-maps the Boolean expressions for the lamp logic are derived:

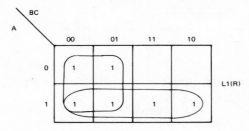

$$L1(R) = A + \bar{B}$$

Similarly (without drawing the K-maps):

$$L1(A) = \bar{A}C$$
$$L1(G) = \bar{A}B\bar{C}$$
$$L2(R) = \bar{A} + \bar{B}$$
$$L2(A) = AC$$
$$L2(G) = AB\bar{C}$$

The traffic light sequencer circuit therefore consists of the modulo-8 counter of Fig. 3.32 plus the combinational logic shown in Fig. 4.7.

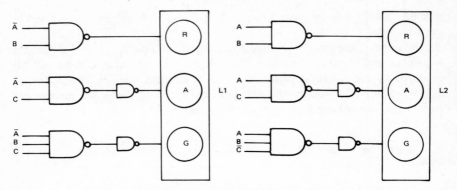

Fig. 4.7 — Lamp driving logic for the traffic lights controller.

Lastly, let us look at the problem of generating a suitable stream of timing pulses, or timing *waveform*. Assume that we have a source of clock pulses with a 1-second interval (say the pulse width is 100 ns). A suitable waveform might have the following time intervals:

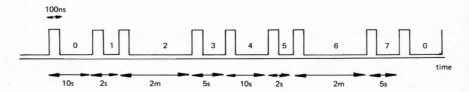

allowing 10 seconds for red-red, 2 minutes for green-red and so on. We have chosen the same timing sequence for both traffic directions, so this gives the basic operating pattern or cycle of:

We have to generate such a pulse interval pattern from the regular 1-second clock source. One solution is to use a 7-bit counter (modulo-128) and divide the cycle up into

where 111 seconds is near enough to 2 minutes for our purposes. Effectively we provide a filter circuit which allows the input clock pulse (of 1-second frequency) through the traffic light controller only at the arrowed points in the above diagram, that is after a count of 10, 12, 123 and 0 in the 7-bit counter.

A suitable circuit diagram is outlined in Fig. 4.8.

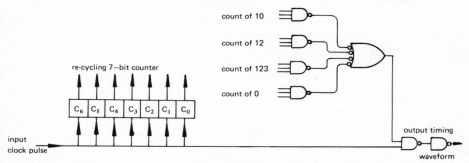

Fig. 4.8 – Circuit diagram of timing waveform generator for traffic lights controller.

4.2 CIRCUIT PROBLEMS IN PRACTICE

In previous sections the point was made that although logic circuits can be designed entirely in terms of abstract operators (AND, OR, NOT) their physical realisation has an important influence on the design process. The wide availability of NAND gates in particular — and the fact that any circuit can be built using only this type of gate — means that circuit designs are usually altered from an AND, OR, NOT form to a NAND (often a NAND, NOT) form. As we have shown, using positive and negative logic conventions, this is a particularly easy conversion to make. It may on occasion be desirable to factorise a Boolean expression to be able to make use of, say, 2-input NAND gates instead of the 3-input type — either because there is a short supply of 3-input gates at the time or more likely to minimise the number of IC packages in a circuit which has some spare 2-input gates from other parts of the circuit. Likewise the use of spare 4-input gates to implement 2- or 3-input functions may be beneficial in terms of gate costs: in the case of NAND gates the unused inputs are simply connected to HI. As we have also seen, NAND gates may be used as inverters in a similar way.

Quite apart from the matching of the circuit design to available components, there are operational problems associated with logic circuits which may not be apparent at the design stage. These problems arise because of the non-ideal performance of circuit components: once again, therefore, they have an impact on the design. The purpose of the present section is to illustrate some of the problems which arise, and how they are solved.

Hazards

Logic gates, flip-flops and circuits of all kinds have inherent *propagation delays*: they take a finite, if small, time to operate. The delay time of a component is characteristic of the speed at which its associated logic family operates. For normal TTL gates and flip-flops propagation delays are about 10 ns and 30 ns respectively.

Consider the operation of an inverter:

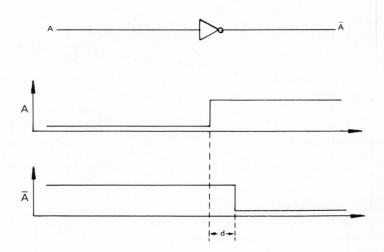

A change at input A causes an inversion at the output $\bar{A}$ a short time later. The propagation delay is shown for d. For time d the law $A\bar{A} = 0$ does not hold true. Likewise the law $A + \bar{A} = 1$ does not hold during the propagation time of the input signal. The violation of these laws of Boolean algebra, even for a very short time, can have serious consequences on the operation of a circuit. The following example demonstrates this.

The Boolean function $F = A\bar{B} + BC$ is implemented by the circuit:

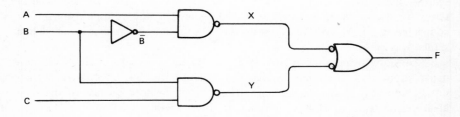

Suppose the initial conditions are $A = B = C = HI$ and that B undergoes a transition from HI to LO. The circuit operation can be represented by the following timing diagrams, where X samples $A\bar{B}$ and Y samples BC (strictly speaking these expressions should be inverted but it is convenient here to use the functional appearance of AND–OR instead of the actual NAND–NAND implementation):

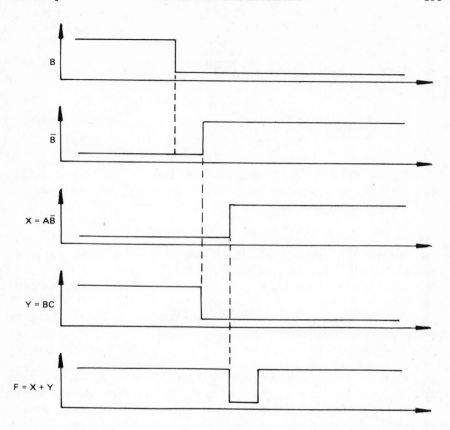

Assuming all the gates have approximately the same propagation delay d, the output F — which theoretically should remain at HI throughout — experiences a negative pulse of duration d. This output *spike* is due to the unequal path lengths for input B in the circuit and to the consequent *race* between the two paths to the output: along one path B experiences two gate delays, along the other a total of three. The resulting output spike is one example of a circuit *hazard*. This particular example illustrates a *single-variable, static* hazard because it involves changes to only one variable and the output remains static (at HI) having undergone a pair of logic transitions.

On its own this circuit malfunction poses no problems, but if F happens to be the input to a sequential circuit it is possible for an incorrect data value to be input to a flip-flop, particularly if F feeds the clock input of an edge-triggered circuit. In practice we want to avoid hazardous operations, but how can this be predicted at the design stage? In the case of a static hazard this can be done by inspecting the K-map for the Boolean expression:

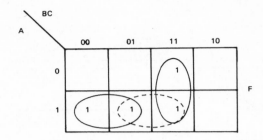

The hazard arises because there are two adjacent 1s not included in the same PI. These are shown by dotted lines round the offending pair. If this (redundant) PI, AC, is included in the final expression, as

$$F = \bar{A} B + B C + A C$$

then hazard-free operation results. This is clearly true if we consider the input conditions A = B = C = HI as before. Since A and C are both HI throughout, while B changes from HI to LO, then AC always remains HI, so F cannot experience an output spike.

Other techniques used in general to eliminate hazards include equalising path lengths by introducing appropriate delays, and avoiding complemented terms (NOT gates) if possible by suitably manipulating the Boolean expression at the design stage.

As well as static hazards, there are two other types of single-variable hazard, called *dynamic* and *essential*. Dynamic types, like the static example above, occur in combinational circuits but there are three different path lengths for the one variable, with inversion in at least one path but not in all three. This type usually arises out of factorisation of the Boolean expression (for example to use 2-input instead of 3-input gates) and can be eliminated by careful re-factorising of the expression. The essential hazard occurs in asynchronous sequential circuits, as a result of a race between an input signal and an internal circuit signal (called a *secondary* signal). Elimination of the hazard can be achieved by inserting delays (non-inverting) in the paths of appropriate signals to equalise the competing path lengths. Lastly there are *multi-variable* hazards which arise when more than one input signal changes at the same time. This type of hazard is generally difficult to eliminate by circuit modification particularly in the case of asynchronous sequential circuits.

Fanout

The data sheets for available logic gates supply a variety of information specifying the physical tolerances of the component, including the supply voltage requirements for successful operation, the power consumed by the gate and the range of ambient temperatures within which it will work. Some of the most important specifications, however, are the low-level and high-level input and output currents which the gate will tolerate. These indicate the *fanout* for the gate, that is the number of other gates (of a similar kind) which may be connected to its output, and also the *load* which this gate imposes on other gate inputs. These facts are extremely important in practice and must be taken into account by the circuit designer.

Let us explain the low- and high-level currents in terms of TTL logic, which is the most popular logic family for gates and other SSI components. The output stage of the standard TTL gate consists, as described in Section 2.4, of two transistors in a totem-pole arrangement. At any time one is ON and the other OFF, or vice-versa, giving the two output voltages LO and HI. The two output cases are illustrated in Fig. 4.9, where (a) shows a HI at the output of gate 1, and (b) a LO voltage.

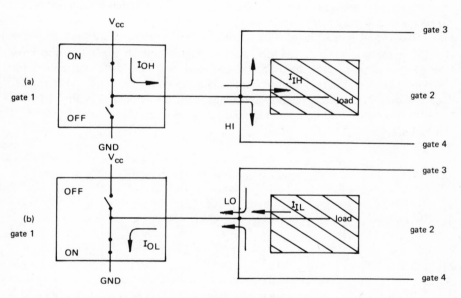

Fig. 4.9 – Schematic diagram of (a) current sourcing and (b) current sinking.

In each case gate 1 is connected to a load (gate 2). Other loads (gates 3, 4 etc.) may be connected as indicated, up to the fanout for gate 1. The two cases, (a) and (b), are referred to as *current-sourcing* and *current-sinking* modes respectively: when the output is at HI, current flows out of gate 1, and when LO it flows into the current gate. In current-sourcing mode if too many loads (represented by I_{IH}) are driven from one gate, the output current of that gate becomes higher than I_{OH} and the output voltage will fall below HI. In current-sinking mode the gate output must be able to sink current I_{IL} from each attached load:

if the total current exceeds I_{OL} the output voltage will no longer be at LO. The current values I_{OH} and I_{OL} are *worst-case* values and determine the fanout of the gate. Values I_{IH} and I_{IL} represent the load which a gate imposes on the output of another. To allow comparisons between different types of TTL gates the manufacturers quote a *standard load* which is:

$$I_{IH} = 40 \, \mu A \text{ (microamps) maximum}$$

$$I_{IL} = 1.6 \, mA \text{ (milliamps) maximum.}$$

A standard TTL gate has a fanout of 10: it can drive 10 standard loads in either a HI or LO state. It must be realised that not all gates represent one standard load nor have the same output characteristics, so when using other than standard gates (for example the increasingly popular low-power Schottky (LS) version — see Fig. 2.18) care must be taken to ensure the fanout is never exceeded otherwise the circuit may malfunction.

The H and S versions of TTL represent larger than standard loads and have a higher fanout, while L and LS varieties impose smaller loads but also have smaller fanouts. In fact the L, S and LS types each have different fanouts for HI and LO cases, but it is safer to quote a *normalised* fanout which is the smaller of the two. Normalised fanouts and input loads are presented in Fig. 4.10 for the five types of TTL component. All figures are quoted in terms of a standard TTL load.

TTL Series	Input load	Fanout
Standard	1	10
H	1.25	12.5
L	0.5	2.5
S	1.25	12.5
LS	0.5	5

Fig. 4.10 — Input load and fanout for the five types of TTL component.

Manufacturers' specifications vary, so it is recommended that data sheets be always consulted — the appropriate data is quoted in terms of the currents, I_{OH}, I_{OL}, I_{IH} and I_{IL}. There are, of course, other logic families but we shall not discuss their fanouts here: this information can be found again from the appropriate data sheets. Note, however, that sometimes it is important to consider the *interfacing* of one logic family with another. The popularity of the TTL family has led to logic interfaces being described as *TTL-compatible* or otherwise. This implies compatibility with TTL voltages ($V_{cc} = +5$ volts, GND $= 0$ volts) as well as with current loadings which would enable a fanout of at least one. The most important of these logic interfaces are MOS–TTL and CMOS–TTL.

Note that although it is convenient to think in terms of only HI and LO voltages there is a specified range of values over which voltages at gate inputs and outputs will be effectively HI or LO. In the data sheets the following values are defined:

V_{IL} is the voltage level required for LO at an input: it is 0.8 volts maximum for TTL;

V_{IH} is the voltage required for a HI at an input; it is 2.0 volts minimum;

V_{OL} is the voltage at a gate output in the LO state: it is guaranteed to be 0.4 volts maximum;

V_{OH} is the voltage at a gate output in the HI state; it is guaranteed to be a minimum of 2.4 volts.

Typical, rather than maximum or minimum values, are 3.6 volts for HI and 0.2 volts for LO. The minimum acceptable voltage for a HI output (2.4V) is 0.4 volts greater than the minimum (2.0V) required to constitute a HI input. Similarly the maximum LO output voltage (0.4V) is 0.4 volts less than the maximum LO input (0.8V). These differences of 0.4 volts are referred to as the high- and low-state *noise margins* of the gate: it is guaranteed to operate successfully when subjected to noise spikes of not more than 0.4 volts amplitude.

Interfacing logic families

While most SSI (and many MSI) components are manufactured in TTL, the majority of LSI components are being made in MOS logic. In particular, microprocessors and associated ICs are mainly MOS devices, and it is often required to interface these to TTL components. Some MOS devices do not have sufficient drive capability for interfacing with TTL: the standard TTL load is too great. However, the L or LS versions present a smaller load and can be so driven. The ease of interfacing depends on the supply voltage of the MOS device in question. If this is a negative value (for example −5V and −12V are common) then interface devices (available in IC form) must be used to translate voltages suitably. If the supply voltage is +5V, which it increasingly is for microprocessors and associated components, the connection is straightforward in terms of voltages. Current loadings are no longer a problem for the newer NMOS devices, and usually one standard TTL load can be interfaced: most NMOS microprocessors and interface devices are designed to be TTL-compatible in this way. Once again, the data sheets should be consulted in specific cases. Older PMOS devices may require the use of an external pull-up resistor (of 1K or less) between the TTL output and +5V when driving MOS from TTL, and an external resistor from the MOS output to GND when driving TTL from MOS. PMOS devices are quite likely to drive only low-power TTL loads.

Increasing use is made of CMOS devices. These may be mixed in a circuit along with TTL components, the CMOS being used where low power consumption is a requirement (but not high speed), and the TTL being used for higher-speed operation. Most CMOS devices have a wide range of supply voltages over which they will operate. For compatibility with TTL a voltage of +5V would be chosen. Some CMOS components will drive a TTL load while others will not. However, special CMOS *buffers* are available which have a fanout of three or four, and which can be used in the interface between CMOS and TTL to improve drive capability. When TTL is used to drive CMOS gates an external pull-up resistor between the TTL output and +5V is recommended. This measure is intended to increase the noise margin at the input to the CMOS gate to an acceptable level.

Schmitt trigger circuits
We have seen how an inverter (or any gate) has a propagation delay between input and output. When the output of a gate changes state, from LO to HI or vice versa, we have idealised the logic transition as an instantaneous change, giving a perfectly vertical slope on a timing diagram. Logic transitions are in practice not instantaneous, but take a finite time, as illustrated below:

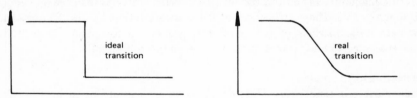

The outputs of TTL gates produce *transition times* of only several nanoseconds and this causes no problems. However, the inputs to TTL gates must change rapidly (transition times of 1 microsecond are unacceptable) otherwise unrelliable operation will result: usually oscillations would be produced at the output of the gate, causing problems if this input is connected to a sequential circuit. If the input to a gate is very slow-changing it can be speeded up — the transition time made shorter — by passing it through a Schmitt trigger circuit, which typically gives transition times of around 10 nanoseconds. Some TTL gates are available with Schmitt trigger circuits built into their inputs. An example is the 7414, a hex inverter package. This type of gate is distinguished by a special shape (indicating a pair of superimposed idealised logic transitions) inside the normal logic symbol as for example:

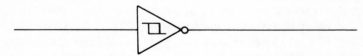

for the 7414 gate.

Monostables

Using the Schmitt trigger circuit a pulse shape can be improved so that its leading and trailing edges have rapid transition times. This can be important for the successful operation of a circuit which requires rapidly-changing inputs. Another important parameter of a pulse is its width or duration. Sometimes the width requires to be altered — lengthened or perhaps shortened. An example where a pulse has to be lengthened is in driving a paper tape reader stepping motor from a logic circuit: the stepping pulse may be required to be as long as 20 or 30 microseconds. A shorter pulse produced by the logic circuit can be lengthened by the use of a *monostable multivibrator,* usually called simply a monostable or a one-shot. This is a device with one stable and one unstable state: it is triggered by pushing it into the unstable state and it reverts after a time to its stable state, the time being dependent on external resistor and capacitor values. An example of a monostable available in IC form is the 74123, which contains two of the devices. A monostable circuit is illustrated below:

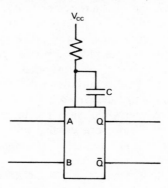

The device has A and B inputs as shown, and can be triggered either by a HI to LO transition at A or a LO to HI transition at B. In the data sheet a graph is given showing the variation of the output pulse width against the values of R and C, thus enabling R, C to be selected for a particular application.

A pair of monostables, such as the pair in the dual 74123 package, can be used to provide a clock pulse without the need for a crystal oscillator circuit. For this purpose they may be connected as follows:

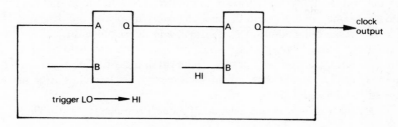

The operation is triggered by a LO to HI transition at one of the B inputs, and thereafter the circuit outputs clock pulses whose width and interval are determined by the R, C values for the two monostables.

4.3 ANALYSIS OF LOGIC CIRCUITS

At least as important as the ability to design logic circuits is the capability of analysing and comprehending circuit diagrams. In the case of the well-practised logic designer this can be taken for granted: synthesising logic circuits is difficult and with practice will lead to the ability to understand other designers' work.

However, for the majority of computer users the opportunity or even the desire to design logic circuits will never arise. It is much more likely that they will be confronted at some time with logic diagrams which have to be understood to enable a hardware fault to be traced or even to clear up a point to allow a program to be written or amended. This type of activity is most likely to occur for those involved with microprocessor work, where often the microprocessor is monitoring or controlling some physical process through a special-purpose hardware interface. Possibly some logic may have to be added to a circuit which is unsuitable for its intended purpose in this specific application: thus in such a case a modest amount of design work may have to be undertaken. The addition of logic to an existing circuit will not in general be possible using a particular design technique, but more usually will be achieved using ad hoc methods: the ability to read and comprehend the original logic diagram is therefore of prime importance.

The choice of material in Chapters 2 to 4 inclusive was motivated partly by the aim of helping the reader to acquire a basic reading knowledge of circuit diagrams. The essentials are the understanding of the various logic symbols and the terminology and labelling used on diagrams and in any accompanying literature. Also important is a readiness to consult manufacturers' data sheets when, inevitably, an unfamiliar IC package designator is encountered.

This is not to say that given an understanding of the material in this book all circuit diagrams will become transparently easy to comprehend. More complex diagrams require time and effort to follow through the signal paths, in order to determine the part which each component plays in the overall circuit operation. The sheer number and density of logic symbols on the diagrams of larger circuits make for confusion to the eye, and consequent difficulties in understanding. It is advisable to try to identify major sub-functions of the circuit and to concentrate on each in turn. Quite often these functions will be separated geographically on the diagram; in such cases boundary lines may usefully be added to help visual discrimination.

An example of a circuit diagram of modest complexity is given in Fig. 4.11.

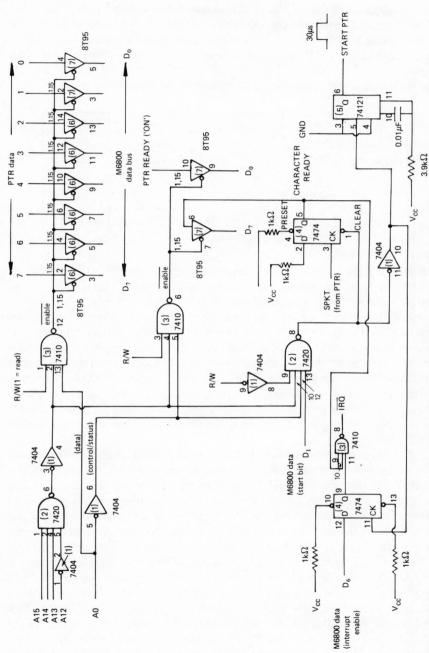

Fig. 4.11 — Microprocessor to paper tape reader interface logic.

This is a diagram of the interface logic for connecting a simple paper tape reader to an 8-bit microprocessor (the Motorola M6800). It is offered as an exercise in attempting to follow the structure and operation of a realistic logic circuit. Some remarks should be made as follows, however, to set the circuit in its context.

The M6800 to reader connection is illustrated by the diagram below:

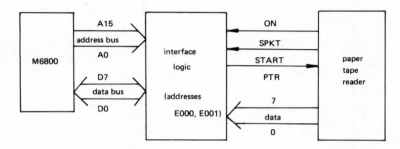

The M6800 has a 16-bit-wide address bus and a bidirectional 8-bit-wide data bus. It employs memory-mapped input/output, which means that any peripheral device is assigned its own address values. Suitable address decoding logic must be provided in the logic interface to recognise these address values when they are generated by the processor. It is common to quote addresses (and data values) in *hexadecimal* (hex for short), that is the base-16 number system with digits 0–9 and A–F (to stand for 10 to 15 inclusive). Each hex digit is represented by four bits. In this case the paper tape reader has two addresses, E000 for exercising *control* and reading *status,* and E001 for inputting *data* from the reader. *Partial address decoding* is used in the interface: only address bits A15–A12, and A0, are decoded. This is possible since no other addresses starting with E are assigned in this particular system.

The paper tape reader has a built-in stepping motor which moves the paper tape on by one frame each time the motor is pulsed. A 30 microsecond pulse is required for this purpose. Each frame or section across the tape has eight data positions — a 1 is represented by a hole, 0 by the absence of a hole — and a *sprocket* hole used for driving the tape by means of a toothed wheel. One step pulse to the motor causes the tape to move so that the next sprocket hole comes to rest over a row of photocells. A light source at the opposite side of the tape allows the binary pattern of the new character to be read — these are available as data bits 0–7. The fact that a new character is ready is conveyed by the SPKT signal which becomes set when a frame is over the photocells. Thus the processor can determine when new data is available by inspecting the CHARACTER READY bit derived from the SPKT signal. This ready bit is passed by the interface logic to the data line D7 of the M6800, but only when the control/status address E000 is generated. Another status bit, PTR READY, is derived from the ON signal which indicates whether the reader is switched on. The processor reads data by reading lines D0–D7 from the data address (E001) of the reader. When a character has been read, the processor can cause the tape to move on by one frame by writing 1 then 0, on data line D1, to the control/status address. The control/status and data addresses may conveniently be regarded as registers with the following layout:

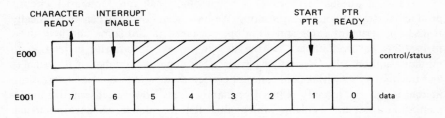

Note that a second control signal, INTERRUPT ENABLE, optionally allows the processor to be interrupted (by means of the $\overline{IRQ}$ line) when a character is ready instead of the *polling* arrangement, whereby the CHARACTER READY line is repeatedly read by the processor.

In Fig. 4.11 the following types of gates are used:

  7404 — hex inverter package

  7410 — triple 3-input NAND gates

  7420 — dual 4-input NAND gates

  7474 — dual edge-triggered D-type flip-flop

 74121 — monostable

  8T95 — hex tri-state buffers

The buffers are used for driving onto a common bus and are tri-state devices. This type of component is represented:

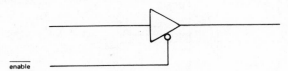

by a (non-inverting) amplifier with the tri-state enable shown entering the side of the symbol. The wiring information — package and pin numbers — is included in the diagram.

4.4 OTHER APPROACHES TO LOGIC IMPLEMENTATION

Logic circuit design has been discussed for both combinational and sequential circuits. In each case it was shown how to design, and implement, the circuits using SSI building blocks — specifically NAND gates and J–K flip-flops. We have thus illustrated components at the bit level, and their use in constructing circuits at the word level. In some of the design examples it was pointed out that the final circuit (or a version of it) is available in MSI form. Word-level units are in turn used to build the processor level in computer logic. Designing computers from this type of building block will be briefly discussed in the next chapter.

The basic design techniques for combinational and sequential circuits produce abstract Boolean expressions. We have seen how they can be physically realised in terms of gates and flip-flops packaged in SSI form, and how these realisations have an impact on the design process, specifically in relation to minimisation criteria. Generally we wish to cut down the number of IC packages and the number of wiring connections between them, so the Boolean expressions in the design are suitably reduced or manipulated. However, alternative implementation tools are available which alter the techniques we have discussed. Instead of using SSI gates and flip-flops it is possible to implement circuits with general-purpose MSI/LSI units effectively containing large numbers of logic elements which can be moulded to suit a particular purpose.

Such units are referred to as *universal logic modules*. The advantage they have over *random logic* — that is, implementation using SSI gates and flip-flops — is their compactness. The packaging of many elements on one IC saves in

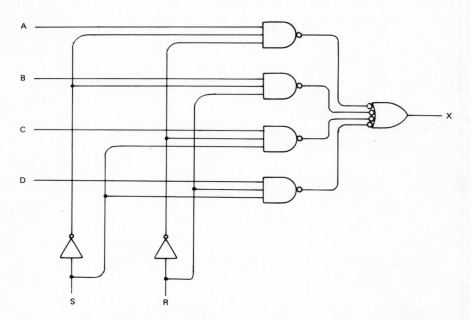

Fig. 4.12 — 1-of-4 multiplexer circuit — A, B, C and D are the data inputs, S and R the control inputs.

space and wiring interconnections. However, universal logic modules are not always more cost-effective, particularly for low-volume applications in which small numbers of identical circuits are to be produced.

Multiplexers

The *multiplexer* (MUX) is a combinational circuit which has several input data lines and a single output data line. Any one of the inputs can be connected to the output at any time, the selection being made by appropriate settings of control inputs. A circuit diagram of a 1-of-4 MUX is shown in Fig. 4.12.
The Boolean equation for the 1-of-4 MUX is:

$$X = A\,\overline{S}\,\overline{R} + B\,\overline{S}\,R + C\,S\,\overline{R} + D\,S\,R \tag{4.1}$$

which is to say that if S and R are both 0 the input line A is selected, and thus effectively connected to X. Similarly if S is 0 and R is 1, B is selected and so on for the other control input combinations. Other MUX configurations deal with larger numbers of input lines (in powers of 2), for example a 1-of-8 MUX has eight data inputs and three control lines.

Multiplexers may be used as universal logic modules for implementing combinational logic functions. Notice that in equation (4.1) all the combinations of S and R are represented. If combinations of 0s and 1s are applied appropriately at the inputs A, B, C and D any of the Boolean functions of 2 variables can be generated at the output X. In fact the circuit is capable of producing all the Boolean functions of 3 variables — consider how to apply combinations of 0s and 1s to two of the data inputs together with combinations of 3 variables applied to the two control inputs and the other data inputs. In general a 1-of-2^n MUX can generate all the Boolean functions of $n + 1$ variables. To implement functions of larger numbers of variables a hierarchy of MUXs may be used. Thus any combinational logic function produced at the design stage may be implemented using MUXs by making the appropriate input connections. Multi-output functions require the use of MUXs arranged in parallel.

Despite the apparent usefulness of MUXs as universal logic modules their cost hardly justifies using them for low-complexity circuits, while for circuits with large numbers of variables there are better alternatives than either MUXs or random logic.

Read-only memories

The *read-only-memory* (ROM) was introduced in Chapter 2, and its variations the PROM and EPROM described. Normally ROMs are used in computer memories to store fixed programs or data. However, another use of the ROM is to realise combinational or sequential logic circuits.

ROMs are available in a variety of configurations. One example, the 32 × 8 ROM, is illustrated in Fig. 4.13.

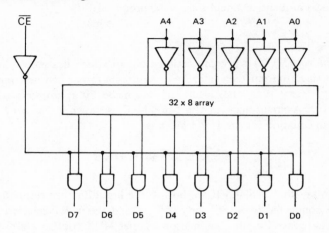

Fig. 4.13 – A 32 × 8 ROM.

There are five address inputs A0–A4, and a chip enable input $\overline{CE}$ which allows data to be read out on the eight data lines D0–D7. Address decoding logic is included on the chip to select one of the 32 stored words of data for output. The contents of the ROM may be viewed as follows:

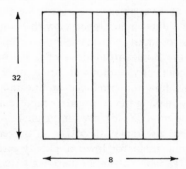

This view shows eight columns each of 32 bits. The columns may each be used to implement directly the output column of a five-variable truth table. The input variable combinations correspond to the 32 different addresses which may be generated by A0–A4. Thus the 32 × 8 ROM can be used to implement any 5-input combinational circuit with (up to) 8 outputs, simply by setting appropriate 1s and 0s into the ROM. If only single output functions are to be realised, however, much of the area in this ROM configuration would be wasted and thus the ROM would be an expensive implementation tool. For large combinational networks the ROM approach may well be cost-effective, particularly because it avoids the wiring interconnections required in a conventional SSI implementation. Note that no minimisation is possible since there is a bit in the ROM for every canonical product term. This is wasteful of space particularly in the case of functions with few product terms (thus leading to a ROM sparsely populated with 1s) but on the other hand no circuit hazards can occur since these are a result of the minimisation process.

Asynchronous sequential circuits can be readily implemented using ROMs: feedback connections are made simply by linking appropriate ROM outputs to input address lines. Synchronous sequential circuits are not so easily realised, and external flip-flops (J-K or D-type) may have to be added to store the values to be fed back to the inputs. The flip-flops are all connected to a common clock: in this way circuit charges can be properly synchronised. This arrangement has the appearance of the Moore/Mealy model of Fig. 3.11.

Programmable logic arrays

The *programmable logic array* (PLA) is available in IC form. It is a general-purpose device for implementing combinational logic functions by means of a 2-level circuit — the familiar sum-of-products form.

The PLA effectively consists of a matrix of wires as shown in Fig. 4.14. There are x rows and y columns, and at the intersection of each row and column the presence of an optional transistor links the horizontal and vertical lines.

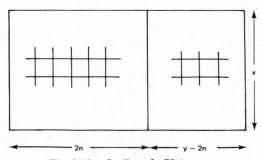

Fig. 4.14 — Outline of a PLA.

An n-variable Boolean function can be realised as follows: $2n$ of the columns are selected and to them each variable and its complement are connected as inputs. The links in the $2n$ by x matrix act as AND gates, while those in the $(y-2n)$ by x matrix behave like OR gates. The required sum-of-products expression results from the appropriate inclusion or exclusion of links. By this means, $(y-2n)$ functions of n variables can be implemented, as long as they each contain at most x product terms.

Unlike ROMs, PLAs can be used to implement minimised expressions because they do not map truth tables directly but have a structure which can be adapted to the (minimised) sum-of-products expression to be implemented. PLAs, like ROMs, are available in field-programmable or factory-programmable versions. A typical size of PLA is one with 16 input variables, 48 possible product terms and 8 function outputs. For field-programmable versions the manufacturers' data sheets contain the information required to program the links.

In a similar way to ROMs, PLAs can be used to implement sequential circuits. If a clock input is provided to transfer data to and from a PLA, it is possible to implement a synchronous circuit using direct feedback paths (no external flip-flops) as long as the delay time in the paths is short compared to the clock period, thus allowing adequate settling of signal values.

Uncommitted logic arrays

Complex logic circuits are sometimes implemented in *custom LSI* logic, whereby an IC is specially designed and fabricated for a customer. The process involves designing several interconnection patterns which will each be added to the silicon chip as a separate layer in fabrication. This design process is very expensive. Once the design is complete, ICs can be mass-manufactured at a very low unit cost. The use of custom LSI therefore tends to be restricted to high-volume applications in which the customer can recoup the high chip development costs.

The idea of the *uncommitted logic array* (ULA) is that a general-purpose IC is fabricated up to the last interconnection layer. The ULA contains a large number of identical *cells* which may be modified by the last fabrication layer to produce a variety of types of logic circuit. The design costs are much reduced since only one layer is involved, therefore the ULA is more suitable than custom logic for low-volume applications.

One of the consequences of using a general-purpose cell is that space is wasted on the chip. Higher circuit packing densities can be achieved by custom design, which attempts to optimise the layout of components on the chip for the particular problem in hand. This implies smaller chip sizes and inevitably lower unit production costs for the custom-designed IC. For high-volume applications, therefore, the ULA approach is less cost-effective than custom design.

The basic cell is a gate-type device (a NAND or NOR element) which can be combined with neighbouring cells to produce a variety of circuits including flip-flops, counters and shift registers. The number of cells required for each circuit depends on its complexity but would typically be four/five for a flip-flop or three/four for each bit in a counter. ULAs typically contain a total of two hundred cells. When the final layer has been added to form the interconnections, the chip is encapsulated in an IC package, for example a DIL package, with a suitable number of pins.

Microprocessors

The use of universal logic modules is part of a trend towards compact general-purpose chips which may be tailored to suit a particular application. This trend has been encouraged by advances in IC technology, particularly by the very high packing densities achieved in LSI. The logic implementation tools described above — MUXs, ROMs, PLAs and ULAs — are useful alternatives to random logic for low-volume applications, where they may be more cost-effective. Random logic, however, is faster in general and may be preferred where speed is of prime concern. High-volume applications of medium to high complexity tend to be dominated either by custom LSI logic, or increasingly by *microprocessors*.

The microprocessor, in contrast to the above *hard-wired logic* units, is a general-purpose device which can be programmed to suit the needs of a customer. Programs, unlike custom logic, can easily be changed so where flexibility is required the microprocessor is preferable. The unit cost of microprocessors is very small but as for custom LSI high design and development costs may have to be borne (for software instead of hardware): the greater the complexity of the task, the higher the costs. In terms of operating speeds a custom LSI chip will probably be better, although faster versions of microprocessors are likely to narrow the gap. Microprocessors are now being used very widely in digital applications previously dominated by hard-wired logic. Even in very low-volume applications the microprocessor may be preferred because of its attractive property of flexibility.

Microprocessors themselves do not constitute a complete logic module. They are basically CPUs which require the addition of a system clock, memory for programs and data — usually a mixture of ROM and RAM — and input/output interfaces which allow communication with the system being monitored or controlled. Thus a complete microcomputer typically consists of not one but several ICs: a computer on a circuit board, in effect. The latest trend in general-purpose devices is that of *single-chip microcomputers* with all the components contained in one IC. Usually this type of microcomputer has an 8-bit CPU, with integral clock, a few hundred bytes (8-bit words) of RAM, a few thousand bytes of ROM (or EPROM) and twenty to thirty single-bit lines for input/output. All that requires to be added is a (5-volt) power supply. These single-chip devices can be used for low-to-medium complexity tasks. Typical application areas include control functions in motor cars, domestic appliances and gaming machines. Higher complexity problems require more storage for programs, and possibly for data, so external memory must be added. As IC technology improves, the capabilities of single-chip computers will be extended by faster processors, larger memories and more input/output lines: the trend continues.

CHAPTER 5

Computer logic design

5.1 COMPUTER LOGIC CIRCUITS

In Chapter 1 (Fig. 1.6) the structural layers of computer logic were outlined. The subsequent three chapters have dealt successively with the following layers: the parallel streams of Boolean algebra and physical realisations; the nature of gates and flip-flops; and their use in designing and implementing logic circuits. We have now reached the stage of describing the final level — how logic circuits are used in constructing computers.

The basic structure of a computer is illustrated by the von Neumann machine of Fig. 1.4. This will be used as a model for the present chapter. The von Neumann machine has three main parts: the CPU, memory and input/output unit. We shall concentrate mainly on the CPU and regard the memory and input/output unit as peripheral devices connected to the processor by means of the address and data buses.

Memory (RAM and ROM) is available in an IC form, as described in Chapter 2, and contains no logic of interest other than the address decoding circuits necessary to locate specific words in the memory matrix.

Choice of memory for a computer system depends on the word length of the computer and on the read/write timing cycles generated by the CPU. The maximum amount of memory depends on the width of the address bus — but also on the requirements of a particular application: in some cases the entire memory address space is not filled because a subset of the total memory is sufficient. A mixture of RAM and ROM is commonly used, especially in microcomputer systems, so that programs (in ROM) need not be reloaded whenever the computer is powered on. Indeed the bootstrap program which is responsible for initiating the loading of other programs from magnetic disc or tape is almost universally stored in ROM and is automatically entered when power is applied to the CPU.

Input/output units are connected to the CPU by means of I/O interfaces, either special ICs or printed circuit boards with the appropriate interfacing logic. Interfaces are either serial or parallel in nature, serial being used for slower peripheral devices like visual display units (VDUs), teletypes and slow paper tape readers, while parallel are employed for fast devices like magnetic discs and tapes, and line printers. Standard interface ICs are very popularly used with microprocessors: the serial types are called ACIAs (asynchronous communications interface adapters) and the parallel ones PIAs (peripheral interface adaptors). These are generally available as part of the family of components for a particular microprocessor and tend to have associated IC designators. For example the Motorola M6800 microprocessor has as relatives the M6820 PIA and the M6850 ACIA. A very simple example of interface logic was given in Chapter 4 (Fig. 4.11) for a paper tape reader to microprocessor connection. This interface logic can in fact be rather easily implemented using an M6820 PIA.

The broad structure of a CPU
Computer architectures are many and various, but are often classified roughly under three headings:

> mainframes
>
> minicomputers
>
> microcomputers.

The differences between these classes are becoming difficult to distinguish as microcomputers and minicomputers both become more powerful, but the distinctions are mainly connected with the richness of hardware and software features which they offer. Mainframes have long word lengths, as mentioned in Chapter 1, with the consequent ability to address more memory and to have a larger repertoire of machine-code instructions. Standard mainframe hardware features tend to include virtual and cache memory to improve the CPU performance. Minicomputers, with a typical word length of 16 bits, offer a more modest set of instructions and have smaller address spaces, but are more often than not equipped with some form of virtual memory, and increasingly with cache as well. Microcomputers are characterised partly by short (8-bit) word lengths but mainly by the fact that the CPU (the microprocessor itself) is entirely contained on one chip. The problems of circuit packing density limit the amount of hardware which can be packaged into a single IC, hence the earlier microprocessors had a very restricted instruction repertoire, few registers and slow performance. Advances in technology now permit 16-bit microprocessors to be fabricated which are at least as powerful as the CPUs of many minicomputers. Thus the distinction between mini and micro is tending to disappear.

Whether a computer is classified as a mainframe, mini or micro it has an architecture which in detail depends on a variety of factors — word length, instruction repertoire, number and type of registers amongst others. Some computers, for example, have multiply and divide instructions — others do not —

while many have a register called a stack pointer used for implementing interrupt handler software calls and subroutines. Nevertheless the majority of computers share a broadly similar CPU structure, as illustrated by our generalised von Neumann machine. In order to discuss the use of logic circuits in computers we look more closely at the black-box units in the CPU. Fig. 5.1 reproduces the CPU part of our von Neumann machine.

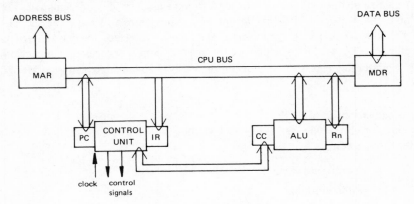

Fig. 5.1 – Outline CPU structure.

The CPU bus is simply a parallel group of wires along which data can be transferred, in either direction, between registers. The number of parallel wires is the same as the word length of the CPU (for example 8 or 16 bits). The important part of the bus is its connection with each register. This is achieved by means of a *bus buffer* (not shown in Fig. 5.1) placed between the bus and the register. The buffer is a bi-directional device with *tri-state outputs* for allowing several devices to be driven on to a common bus.

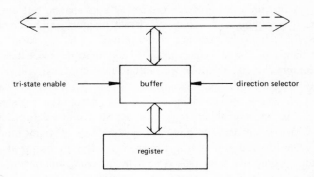

A similar uni-directional device, the 8T95 driver, was described in Section 4.3. Bus buffers, otherwise called *transceivers,* are available as octal IC packages. An example is the 74245 (or its low-power Schottky equivalent, the 74LS245) illustrated below:

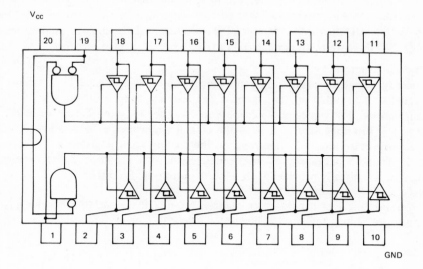

The eight non-inverting buffers for each direction have a common tri-state enable line (active-LO). A direction selector line chooses the direction of data transfer. Note that the tri-state property is used only in the direction of transfer to the bus. It is not required for connection between the buffer and the (single) register. This chip would be used also for the external data bus to the memory and input/output units. An octal bus driver, the 74244, is available for use in driving the external address bus from the MAR register.

The control unit is responsible for sequencing all the actions within the CPU, deriving its basic timing from a system clock which is usually a crystal oscillator circuit external to the CPU. As described in Chapter 1, a computer fetches machine-code instructions stored in memory, decodes and obeys them one by one. These actions correspond to micro-instructions which are mostly register transfers — the transfer of data from one register to another across the CPU bus. The order in which machine-code instructions are obeyed is determined by the program itself. Unless the present instruction is a JUMP or conditional BRANCH the next instruction to be obeyed is the one stored immediately after the present one in memory. Each micro-instruction is initiated by control signals issued by the control unit. In particular, register transfers are controlled by the opening of a path between a source register and a destination; this is achieved by enabling the (tri-state) output of the source register and then pulsing the clock inputs of the flip-flops in the destination register to cause data to be transferred across the bus.

In the next section the detailed operation of the control unit will be demonstrated by a simple example. The design of the control unit is one of the most important activities in computer design. Two approaches are used — the hardwired and the microprogramming methods — and these merit further description in the last section of the chapter. The remainder of the present section deals with the other two main units in the CPU — the ALU and the registers.

Register structure

Basically a register consists of a number of flip-flops each capable of storing a single bit of information. The number of flip-flops depends on the size of register required, but typically would be 8 or 16. Edge-triggered D-type or master-slave J-K flip-flops may be used — the former triggering on the leading edge of the clock pulse, the J-K effectively on the trailing edge. Registers are available in IC form, a package typically containing 4 or 8 flip-flops. Some of these registers are already provided with tri-state outputs and can therefore be connected directly onto a bus without the need for a bus buffer.

The transfer of data between registers can be either *serial* (one bit at a time) or *parallel* (all bits together) as illustrated below:

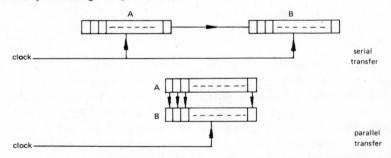

In serial transmission, data is shifted one bit at a time in one direction in both the source (A) and destination (B) register. Thus each register must be internally organised as a *shift register* in which the output of each flip-flop is connected to the input of its neighbour. Computers usually provide machine-code instructions to shift data in general-purpose registers right or left, either transferring the data into another register or shifting it simply within a single register, in which case data bits are lost from one end. *Rotate* instructions move data round from one end of a register to the other and require an end-to-end connection as illustrated:

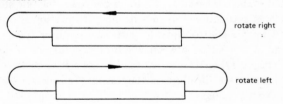

In a shift transfer, each source and destination flip-flop is connected to a common clock source. An n-bit shift operation requires n pulses of the clock. However, only one wire need link the two registers in the transfer.

In contrast a parallel data transfer requires one wire for each register bit. Only the flip-flops in the receiving register need to be connected to the clock source since the flip-flops in the source register are not themselves receiving new data. The transfer can be accomplished with one pulse of the clock.

A section of a general-purpose register based on J–K flip-flops is shown in Fig. 5.2.

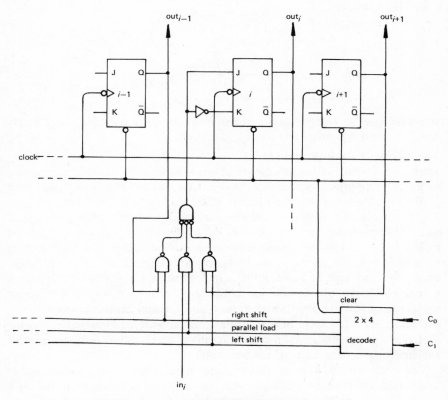

Fig. 5.2 – Section of general-purpose register.

This shows the logic required for bit i of the register, an effective AND–OR network (shown using NAND gates) with inputs from the Q output of flip-flops $i - 1$ and $i + 1$, and an input labelled in_i to which can be connected the i-th output of any other flip-flop; each bit of the register requires an identical AND–OR network. The outputs of all the flip-flops are available in parallel for transfer to another register. Common to the whole register is a *2 X 4 decoder* by means of which the required register function can be selected, using the four combinations of inputs C_0 and C_1. Four functions are provided: left shift, right shift, parallel load and clear. A decoder is a de-multiplexer (a MUX in reverse) without an input data line — the selected output line has a 1 on it, the other three are set at zero. Alternatively the decoder can be considered as a de-MUX with an input line set permanently at 1.

Note that the decoder outputs act as *enable* inputs to the 2-input NAND gates, allowing the data on the other input through to the 3-input NAND gate if the enable line is at 1, otherwise disabling the output of the 2-input gate. This is illustrated below:

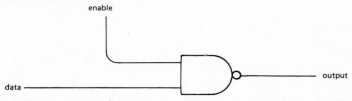

If enable is set to 1, input data is transferred (inverted) to the output; if enable is 0 the output is permanently set at 1 (disabled).

The arithmetic and logic unit (ALU)

Machine-code instructions specify operations to be carried out on data contained in the general-purpose registers of the computer, and in some cases on data held in memory. These operations are of two kinds: arithmetic and logical.

The arithmetic operations are of course addition, subtraction, multiplication and division. Most microprocessors and some minicomputers do not provide multiplication and division, while mainframes have special-purpose logic circuits to provide these functions. In machines without hardware multiply and divide, these functions must be provided by software routines which basically perform a repetitive series of additions or subtractions. The more advanced hardware units feature *floating-point* multiplication and division in which operands and and result are represented in a scientific notation with mantissa and exponent. This type of operation is not necessary for many text-handling or simple control applications so a more basic ALU is adequate.

In computers it is usual to represent numbers using the *2s complement* notation. In this system a number is stored in an n-bit register with the least significant $n - 1$-bits representing the magnitude of the number and the most significant bit its sign — whether it is a positive or a negative number:

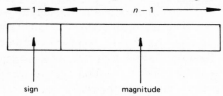

Positive numbers have a sign bit of value 0 and a magnitude in true binary form, for example (considering a 4-bit register for simplicity) the decimal number +5 is stored as follows:

0	1	0	1	+5

Negative numbers are derived from positive numbers in the following way: each bit, including the sign bit, is inverted (complemented) then 1 is added to the least significant end. Thus −5 is represented as:

1	0	1	1	−5

The magnitude is not in a true binary form for negative numbers. Note that the sign bit automatically becomes 1, indicating that the magnitude is a negative quantity. The advantage of storing numbers in this way is that arithmetic becomes very simple: all numbers are added together, and the signs and magnitudes look after themselves. Consider the operation:

$$5 - 3$$

Using 2s complement arithmetic this is effectively

$$5 + (-3)$$

where (−3) is the 2s complement negative representation of 3:

1	1	0	1	−3

The arithmetic operation $5 - 3$ is then an addition process:

0	1	0	1		+		1	1	0	1
	+5							−3		

giving the result:

0	0	1	0	+2

which is what we would expect. Notice that positive 2s complement numbers can be derived from their negatives by exactly the same process used in converting positive forms to negative: invert all the bits and add 1.

The basis for all four arithmetic operations is the binary full-adder (described in Chapter 3). Using 2s complement numbers subtraction is effectively achieved by the addition process. Multiplication is basically a series of repeated addition (and shifts) as illustrated in Fig. 5.3.

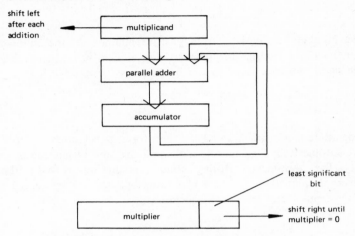

Fig. 5.3 — Illustration of multiplication by repeated additions.

The two operands are contained in registers labelled *multiplicand* and *multiplier*. These are to be multiplied together to give a result in the *accumulator* register which is initially cleared to zero. Binary multiplication is basically very simple: if the least significant bit of the multiplier is 1, the multiplicand is added into the accumulator; if it is 0, no addition is performed. Then the multiplier is shifted one place to the right, and the multiplicand one place to the left. If the contents of the multiplier register is now zero the multiplication is complete. Otherwise the process is repeated: the least significant multiplier bit is checked for 1 or 0, and the next addition, if any, performed. This method assumes that the operands have both been converted to a positive form. The sign of the result can be obtained by checking the original signs of the two operands. A slow hardware multiplication unit can be based on the binary adder using the above method. Division can be provided similarly by an iterative method consisting of subtractions and shifts, but the (2s complement) subtractions performed by a binary adder.

Logical operations in computers are required to manipulate the bits held in registers (or in memory). They include shift operations (left, right, rotate) whereby data is moved within a register; NOT, in which all bits of an operand are inverted; and AND, OR, NOT between pairs of registers, the appropriate logical operation being performed between corresponding bits in the two registers. Logical operations, too, can be based on a binary adder, as we shall now see.

Detailed structure of an ALU

In Chapter 3 (Fig. 3.15) the circuit for a binary full-adder was derived. The adder takes two single-bit operands and a carry-in, and provides a sum bit and a carry-out. A simple ALU can be constructed from the basis of a binary full-adder circuit, one circuit for each bit in the word length of the computer. Fig. 5.4 shows a single-bit stage of such an ALU. The binary full-adder circuit is enclosed in dashed lines with the inputs and outputs of Fig. 3.15 in brackets.

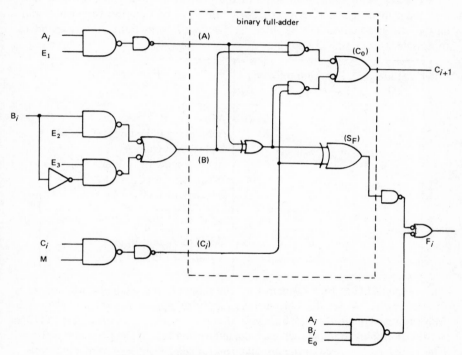

Fig. 5.4 – Single-bit stage of a simple ALU.

The ALU circuit takes inputs A_i and B_i from bit i of the two operand registers and a carry-in C_i from the previous (less significant) stage of the ALU. It produces a *function output* F_i and carry-out C_{i+1} which goes to the next, more significant stage. The full-adder circuit has been extended with extra logic. The *mode control* M and *function selectors* E_0-E_3 are inputs which enable the circuit to produce a number of alternative functions of the operands. When the mode control input is at 1, arithmetic operations are provided at F_i by selecting combinations of the function selectors. The mode control input set to 0 gives a choice of logical functions instead. In the former case, the (2s complement) arithmetic operations add (ADD) and subtract (SUB) make use of the carry input and output bits. For all other operations carry is ignored (except for the least significant carry in C_1).

Note that for subtraction the least significant carry in bit (C_1) will be set to 1: together with the inversion of B using input E_3 this has the effect of forming the negative 2s complement of the B operand. Similarly an increment operation (INC) can be implemented by setting $C_1 = 1$ and selecting only one operand, A or B.

The following table gives some of the possible combinations of mode control and function selectors together with the type of function for each combination. In subtraction (SUB), bits are *borrowed* from rather than carried to successive stages in the operation.

Mode	Function selectors				Function
M	E_0	E_1	E_2	E_3	
1	0	1	1	0	ADD A to B (with carry in and out) $(C_1 = 0)$
1	0	1	0	1	SUBtract B from A (with borrow in and out) $(C_1 = 1)$
1	0	1	0	0	INCrement A $(C_1 = 1)$
1	0	0	1	0	INCrement B $(C_1 = 1)$
0	0	1	1	0	EOR of A_i and B_i
0	1	0	0	0	AND of A_i and B_i
0	1	1	1	0	OR of A_i and B_i

The simple ALU in Fig. 5.4 is in turn the basis of an available 4-bit-wide ALU IC. Called the 74181, this chip has the functional appearance shown in Fig. 5.5. Most computers are 8 or 16 bits wide therefore in practice two or four 74181s would be employed in parallel, each dealing with 4 bits of the processor registers. The 74181 ALU is used in the popular 16-bit PDP-11 range of minicomputers.

Most of the input and output connections in Fig. 5.5 are self-explanatory. However, A = B is a *comparator* output giving 1 if each $A_i = B_i$ (otherwise giving 0). P, G, C_i and C_{i+4} are used for linking a 4-bit ALU to neighbouring ALUs (for example in a 16-bit computer) to improve the overall speed of arithmetic operations: they are called *carry look-ahead* connections.

Carry look-ahead

The use of a parallel binary adder in an ALU involves the generation of carry bits for each successive stage of the operation. In effect the addition is not truly a parallel operation because the correct carry-in bit is not available as soon for the more significant end as it is for the less significant end of the adder. At each stage of addition the carry output is produced by 2-level circuit, indicated by gates 1, 2 and 3 in Fig. 5.6.

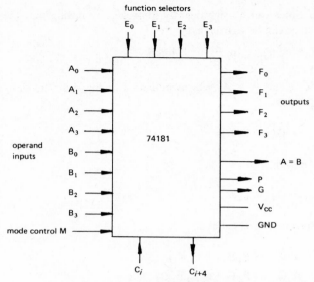

Fig. 5.5 — Functional outline of 74181 ALU.

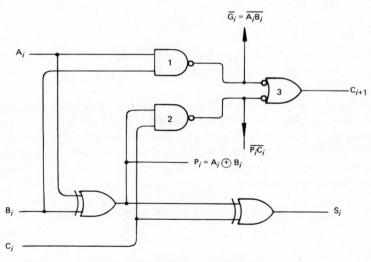

Fig. 5.6 — The ith stage of a parallel adder.

Assuming a 10 nanosecond delay per gate, the carry-out is produced for each stage after 20 nanoseconds, once the carry-in from the previous stage is available. For an adder with n stages, the most significant carry-in bit is available after $20 \times (n - 1)$ nanoseconds, which for a 16-bit addition gives 300 ns in total.

Addition times can be significantly improved by noting that the carry-out from the ith stage can be written as:

$$C_{i+1} = G_i + P_i C_i$$

where $G_i = A_i B_i$ and $P_i = A_i \oplus B_i$ as indicated in Fig. 5.6. The carry-out from the first stage is thus

$$C_2 = G_1 + P_1 C_1$$

and from the second

$$C_3 = G_2 + P_2 C_2$$

which can in turn be expressed as

$$C_3 = G_2 + P_2 (G_1 + P_1 C_1)$$
$$= G_2 + P_2 G_1 + P_2 P_1 C_1$$

The third-stage carry output is

$$C_4 = G_3 + P_3 C_3$$
$$= G_3 + P_3 (G_2 + P_2 G_1 + P_2 P_1 C_1)$$
$$= G_3 + P_3 G_2 + P_3 P_2 G_1 + P_3 P_2 P_1 C_1$$

Each successive carry-out can be expressed in terms of Ps, Gs and C_1 only, in a 2-level sum-of-products form. Each P_i and G_i depend only on A_i and B_i. Therefore every carry-out bit may be produced from values available at the start of the addition process, in a truly parallel operation. P_i and G_i are referred to as *propagate* and *generate* terms respectively. In general the carry-out from stage i can be expressed using propagate and generate terms:

$$C_{i+1} = (G_i + P_i G_{i-1} + P_i P_{i-1} G_{i-2} + \ldots + P_i P_{i-1} \ldots P_2 G_1)$$
$$+ P_i P_{i-1} \ldots P_1 C_1$$

This expression becomes more complex the larger the value of i, the complexity being in the number and size of the product terms. In practice, *carry look-ahead* logic circuits based on the above general expression are available for 4-bit groups only. Carry look-ahead is provided in the 74181 ALU as indicated by the P, G, C_i and C_{i+4} connections in Fig. 5.5.

Addition times over groups of 74181s can be improved by the use of an external carry look-ahead generator. Such a generator is available as IC 74182, providing for connection of up to 4 74181s. Its functional appearance is shown in Fig. 5.7.

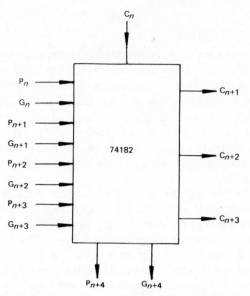

Fig. 5.7 — Functional outline of 74182 carry look-ahead generator.

The use of an external carry generator is illustrated as follows for a 16-bit ALU:

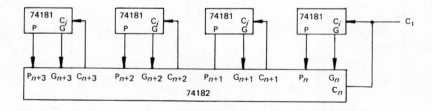

The C_{i+4} output from each 74181 and the P_{n+4}, G_{n+4} outputs from the 74182 are not used in this configuration.

A look-ahead generator typically adds 10 ns to the basic carry-production time of a single stage, but this time is the same for all 4 bits of a 74181. If four 74181s are connected without an external 74182 there are 3 carry production times of 30 ns each before the most significant ALU has the correct carry-in available. This gives a total delay of 90 ns, a significant improvement on the 300 ns of the bare 16-bit adder (without any carry look-ahead logic). With the arrangement shown above, a 74182 linking 4 74181 ALUs, the total time for a

16-bit addition is about 40 ns, a factor of around 8 times better than the bare adder.

5.2 CONTROL STRUCTURE

The fetching and obeying of machine-code instructions corresponds, as we have seen, to sequences of micro-operations within the CPU. The initiation of micro-operations, and their correct sequencing, is the responsibility of the control unit. It derives its basic timing from a clock source and controls the operations by means of a set of control signals. The design of the *control structure* (the control unit and its associated control lines) is of central importance to the architecture of a computer, since it determines the micro-operations which can be provided and consequently the set of machine-code instructions available for use by the programmer.

To give the flavour of a control structure we now consider a simple example based on the outline CPU shown in Fig. 5.1. A more detailed version of the CPU is illustrated in Fig. 5.8 with control signals added, including the ALU mode control M and function selector inputs E_0-E_3. The number of general-purpose registers Rn has been chosen as two: R0 and R1. These are shown with a 1-of-2 MUX connection into the ALU. The MUX goes into the B inputs of the ALU, while the bus is connected to the A inputs and F outputs.

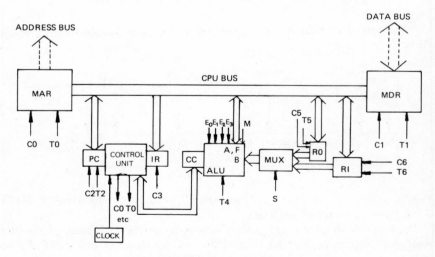

Fig. 5.8 – CPU showing control signals.

Selection of one of the two registers is determined by the value of the control input S: 0, 1 selects R0, R1 respectively. Apart from S, M, and E_0-E_3, all the

C_i and T_i are outputs of the control unit (not all shown explicitly in the diagram). Each C_i is connected to the clock input of one of the registers in the ALU. When data is to be transferred into a register the appropriate C_i line is pulsed by the control unit. All transfers are in parallel mode.

The T_i are tri-state enable lines used to control the connection of registers onto the CPU bus. Since IR does not have to output to the bus it has no tri-state enable line. The ALU, being a combinational circuit, has no C_i (clock) line but is connected to the bus by means of a bus buffer which has input T_4 for allowing data to be transferred onto the bus. Although the registers (apart from CC and IR) and ALU have bi-directional communication with the CPU bus, and corresponding direction selection lines which are set/reset by the control unit, these control lines are not shown in the diagram. We shall assume also that each T_i is an active–HI line, in other words when T_i is 1 data can be output; if it is 0 the outputs are in the third (high-impedance) state and are effectively isolated from the bus. When data is to be transferred onto the bus the appropriate T_i is set to 1 (all others will be set at 0) by the control unit.

Thus in a register transfer operation, over the CPU bus, one T_i is set to 1 and one C_i is pulsed. This corresponds to setting up a path between the source and destination, as illustrated below for the transfer

PC $\rightarrow$ MAR

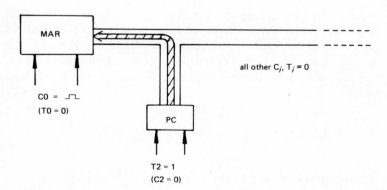

The relative timing of the signals is important. Source data (from PC) must be available on the bus when the destination register (MAR) is pulsed. This is possible as long as T_2 is set to 1 before or during the clock pulse applied to C_0: assuming J–K flip-flops are used to implement the registers, the trailing edge of the pulse must occur after the outputs of the PC register have propagated onto the bus and are available at the MAR inputs. T_2 must be maintained at 1 until after the trailing edge of the pulse. The control unit can sequence these actions in relation to a pulse derived from the clock source as follows:

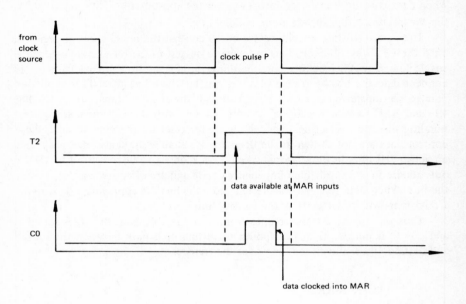

In response to the leading edge of clock pulse P, T_2 is set to 1. After a suitable propagation delay data is available at the MAR inputs. C_0 is pulsed so that the data is clocked into the MAR register, then T_2 is reset to 0.

Let us illustrate the detailed operation of each register transfer in a typical machine-code instruction. We use the example given in Chapter 1, using R0 specifically:

ADD R0, X

('add the contents of memory location X to register R0').

Fig. 5.9 indicates the control line settings appropriate to each register transfer. Control lines for the memory accesses are not shown (using the external address and data buses). 'Decode IR' is an operation internal to the control unit: we shall discuss this operation shortly. Don't care conditions are indicated with Xs, and the pulsing of a C input (while the corresponding T is at 1) by ⎍

Phase	Micro-instruction	Control Signals																	
		C_0	T_0	C_1	T_1	C_2	T_2	C_3	T_4	C_5	T_5	C_6	T_6	S	M	E_0	E_1	E_2	E_3
fetch	PC → MAR	1	0	0	0	0	1	0	0	0	0	0	0	X	X	X	X	X	X
	(MAR) → MDR*	0	0	0	0	0	0	0	0	0	0	0	0	X	X	X	X	X	X
	MDR → IR	0	0	0	1	0	0	1	0	0	0	0	0	X	X	X	X	X	X
decode	decode IR†	0	0	0	0	0	0	0	0	0	0	0	0	X	X	X	X	X	X
obey	PC + 1 → PC	0	0	0	0	1	0	0	1	0	0	0	0	X	1	0	1	0	0
	PC → MAR	1	0	0	0	0	1	0	0	0	0	0	0	X	X	X	X	X	X
	(MAR) → MDR*	0	0	0	0	0	0	0	0	0	0	0	0	X	X	X	X	X	X
	MDR → MAR	1	0	0	1	0	0	0	0	0	0	0	0	X	X	X	X	X	X
	(MAR) → MDR*	0	0	0	0	0	0	0	0	0	0	0	0	0	X	X	X	X	X
	MDR + R0 → R0	0	0	0	0	0	0	0	1	1	0	0	0	0	1	0	1	1	0
next	PC + 1 → PC	0	0	0	0	1	0	0	1	0	0	0	0	X	1	0	1	0	0

Fig. 5.9 – Control signal settings for ADD R0, X.

*memory access operations.
†internal to control unit.

Note that the carry-in to the least significant bit of the ALU is not shown in Figs. 5.8 and 5.9. The operation

$$PC + 1 \rightarrow PC$$

is implemented using an INCrement operation in the ALU for which the least significant carry-in would be set to 1.

Both operations

$$PC + 1 \rightarrow PC$$

and $$MDR + R0 \rightarrow R0$$

show more than one T line enabled. Within these operations both T_i are not set to 1 at the same time but must be disjoint as shown below for $PC + 1 \rightarrow PC$:

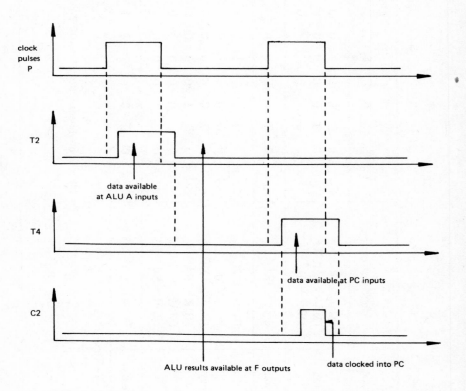

There are effectively two register transfers and this requires two clock pulses P for the operation. The first of the two clock pulses would also initiate the setting of ALU inputs M and E_0-E_3 for the INCrement operation. Similarly for MDR + R0 → R0:

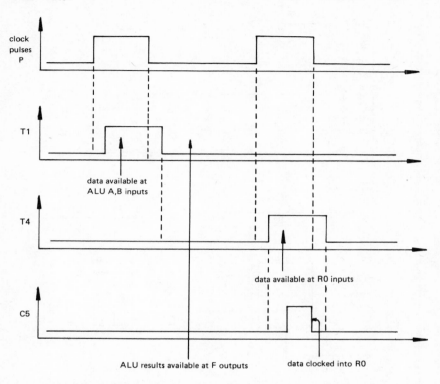

In this case the first clock pulse would initiate the setting of M, E_0-E_3 (for ADDition) and also S, the MUX selector input.

The above register transfers are part of a longer sequence of similar transfers. For a given machine-code instruction the control unit must be able to set/ reset each T_i and pulse each C_i in the correct order. To do this it can use a *sequence counter* which has the functional appearance shown in Fig. 5.10(a). The idea of the sequence counter is that the stream of clock pulses is divided into *cycles,* each cycle consisting of a fixed number of clock pulses. The slowest machine-code operation — the one which requires the largest number of micro-operations — determines the cycle length. The sequence counter in the diagram has a cycle length of n and outputs P_1 to P_n which carry separate *phases* of the clock cycle as indicated in Fig. 5.10(b) for $n = 8$. Each P_i will initiate the operations of a set of control lines for step i of a machine-code instruction. A sequence counter can be constructed either from a modulo-n counter and an n-output decoder, as shown in Fig. 5.10(c) for $n = 8$, or from an n-bit ring counter which requires no decoding logic (but is expensive in flip-flops).

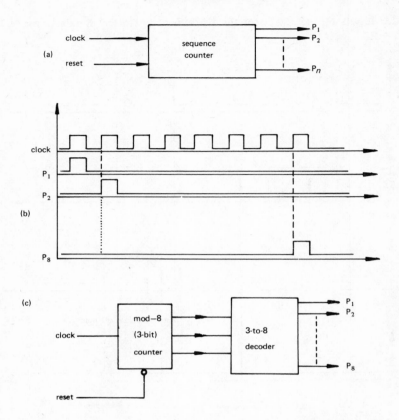

Fig. 5.10 – Sequence counter: (a) block diagram; (b) phased outputs; (c) construction.

Each machine-code instruction has four basic steps in it, each step consisting of one or more micro-instructions as we have seen: of the four steps for each instruction:

(1) fetch

(2) decode

(3) obey

(4) next

(1) and (2) are standard for all instructions but (3) and (4) depend on the instruction contents. Therefore the n phases of operation will differ (after (2)) for all instructions and so each P_i will not in general initiate the same set of control lines for one instruction as it does for another. The information which

determines the control lines to be initiated by each P_i for a particular machine-code instruction comes from the *instruction decoder*. This is a combinational circuit which decodes the contents of the IR as shown below.

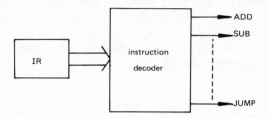

Only one of the outputs is active at any time and distinguishes the current instruction.

The control unit consists, thus far, of a sequence counter and instruction decoder, fed by a clock input and the contents of the instruction register. Its outputs, as we have seen, are the various control lines which allow micro-instructions to be implemented. To implement the entire control structure of the CPU we require further logic in the control unit which maps the instruction decoder outputs and sequence counter phasing signals onto the appropriate control line outputs: we shall call this the *micro-instruction encoder* since the control lines effectively determine the micro-instructions obeyed.

The encoder is basically a multi-input, multi-output combinational logic circuit. In practice a computer may require a few hundred control line outputs, of the order of a hundred machine-code instructions and about ten phases in the instruction cycle, making the design of the control structure a lengthy and expensive process. The general structure of a control unit is outlined in Fig. 5.11.

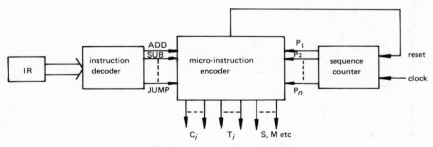

Fig. 5.11 – Outline of a complete control unit.

When a computer is being designed a number of choices will be made before the details of the control structure are tackled. Important dimensions — word length, address bus width, number of registers — would be amongst the first matters to be settled. Ultimately the computer is to support software so the set of machine-code instructions which it will provide is extremely important. These would be specified at an early part of the design process. Using the notation of *register transfers* each machine-code instruction can be expressed as a sequence of micro-instructions (as for the example ADD Rn, X) and thus the required control structure devised. Indeed the overall operation of the CPU can be written in a (pseudo) flow-chart form, specified down to the level of detail of each micro-instruction. A section of such a flow-chart is illustrated in Fig. 5.12 (using our simple example).

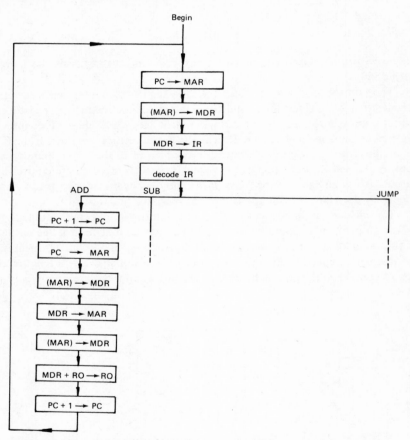

Fig. 5.12 – Micro-instruction flow-chart for overall computer operation.

Once the control signals have been specified and their values written down for each micro-instruction – in a truth-table form similar to that of Fig. 5.9 – the micro-instruction encoder logic can be designed. It could be implemented in discrete combinational logic, using SSI packages, or possibly using ROMs or

PLAs. The implementation of the micro-instruction encoder using combinational logic fixes the machine-code instruction set in hardware and is usually referred to as the *hardwired* method of designing computers. A hardwired control unit does not easily allow the modification of machine-code instructions. An alternative method of implementation which allows the flexibility of changing the machine-code instruction set is that of *microprogramming*.

5.3 MICROPROGRAMMING

Microprogramming was first proposed by M. V. Wilkes in 1951 as a methodical way of designing computers. It uses a *control memory* for storing micro-instructions as a series of 1s and 0s. In the same way as a computer program consists of a sequence of machine-code instructions, each machine-code instruction is represented by a sequence of micro-instructions in a control memory called a *microprogram*.

Whereas in a hardwired control unit the setting and unsetting of control lines for each machine-code instruction is fixed in a combinational logic circuit and is therefore difficult and expensive to alter, a machine-code instruction in a microprogrammed unit may be changed easily and cheaply by amending a microprogram. Since control memory is implemented by a ROM the amending of a microprogram corresponds to replacing or re-programming a ROM chip.

Microprogramming offers a methodical approach to designing computers in the first instance, but also allows for the possibility that the machine-code instruction set is either not fully specified or has design errors in it when the hardware comes to be constructed. The designer or manufacturer can amend or add machine-code instructions later. Originally control memories were all read-only devices. Some modern computers offer a *writeable control memory* (often called a writeable control store or WCS) whereby the machine-code instruction set can be modified by the user. Such a control memory is termed *dynamically microprogrammable* since it can be altered by a program running on the CPU.

Each machine-code instruction is represented by a microprogram in control memory. When the IR for a particular instruction is decoded, this provides the starting address of the corresponding microprogram. Once a microprogram has started being obeyed the sequencing of micro-instructions is specified within the microprogram itself: each micro-instruction consists of a *control field* and an *address field*. The control field contains the settings of the required control signals. These are read into the *control memory data register* (CMDR) when a micro-instruction is accessed. Its address field bits provide inputs for the *control memory address register* (CMAR). Other inputs to the CMAR come from the IR, effectively locating the start adress of the micro program, while the micro-instruction address field gives the address (within the microprogram) of the next micro-instruction to be obeyed. An outline microprogrammed control unit is shown in Fig. 5.13.

The CMAR and CMDR can be considered to operate much as the MAR and MDR of a conventional memory. Included in the control memory (but not shown in the diagram) is address decoding logic to access the word specified in the CMAR. Comparing this microprogrammed control unit with the (hard-wired) unit of Fig. 5.11, we see that the control memory replaces the micro-instruction encoder logic (and its address decoder is equivalent to the instruction decoder). The sequence counter has disappeared. It is effectively replaced by the ability of the micro-programmed unit to obey a sequence of micro-instructions once a micro program has been initiated: the microprogram itself does the sequencing. The clock source shown in Fig. 5.13 controls the rate of fetching micro-instructions from control memory. The diagram implies that all control line outputs from the CMDR become valid at the same time. In practice it is possible to reduce the number of micro-instructions required for any machine-code instruction by enabling sub-groups of the control lines at different times within one micro-instruction clock period. This can be achieved by providing several phased clock inputs — using a sequence counter technique — each input controlling a different segment of the CMDR and hence a sub-group of control lines.

Generally the phases within a micro-instruction period will correspond to the phases in which machine-code instructions themselves are fetched and obeyed. It is possible for the fetch and obey phases to be overlapped: the next instruction is fetched while the current one is being obeyed. This can save considerable amounts of time in running a microprogram. Overlapping instructions may necessitate the use of another register between control memory and the CMDR to hold the next instruction while the previous one is still being interpreted in the CMDR.

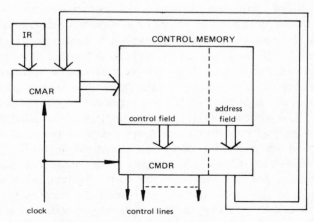

Fig. 5.13 — Outline of a microprogrammed control unit.

Branching

Fig. 5.13 does not show any provision for *conditional branching*, which is used in microprogramming in a way similar to that in machine-code programming. Such branches will be conditional on the value of signals derived from external sources, such as the flip-flops in the CC register associated with the ALU. In order to provide for this type of branching, the 'next address' generation scheme indicated in Fig. 5.13 must be modified. Just as for machine-code instructions it can be assumed that the next instruction is to be fetched from the next memory location unless the present instruction is either a conditional or an *unconditional* branch (an unconditional branch is equivalent to a JUMP instruction). For both types of branch instruction the next address must be specified; for conditional branches the external condition on which the branch depends must also be specified in the micro-instruction. For all other instructions the next address is generated simply by incrementing the CMAR contents. In the case of conditional branches, if the specified condition is not satisfied, the next address in the instruction is ignored and the CMAR incremented as for other instructions. Unconditional branches always cause the CMAR contents to be altered to the new address specified in the instruction.

The *micro-instruction format* thus has a more complex structure than simply a control field. It will more realistically contain the following fields:

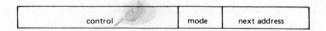

where *mode* may for example be a 3-bit field with the possible values:

(0) 000 = no branch (increment CMAR for next address)

(1) 001 = unconditional branch (use specified next address)

(2) 010 = branch if external condition X1 = 1

(3) 011 =   ,,   ,,   ,,   ,,   X2 = 1

.

.

.

.

(7) 111 = branch if external condition X6 = 1

For the 'no branch' case the next address field is effectively empty. In the 'branch if' cases the specified next address is used if $X_i = 1$ otherwise the CMAR is incremented.

This scheme could be implemented by modifying the simple microprogrammed control unit (of Fig. 5.13) as indicated in Fig. 5.14.

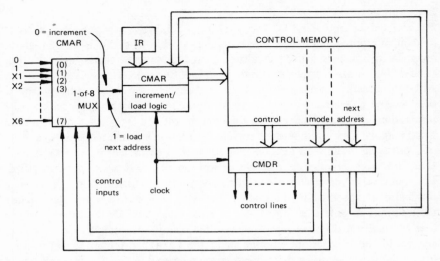

Fig. 5.14 — More detailed microprogrammed control unit.

The control memory now has three fields. The mode field provides control inputs for a 1-of-8 MUX which in turn initiates logic either to increment the CMAR or to load it with the next address field of the CMDR. MUX data inputs (0) and (1) are wired permanently to 0 (GND) and 1 (HI) respectively, the other six to external flip-flops Xi. Either mode (0), or any of the modes (2) to (7) when $Xi = 0$, causes the MUX output to be 0, thus the CMAR is incremented. If the mode field is either (1), or (2) to (7) when $Xi = 1$, the MUX output is 1 and then the next address field is loaded into the CMAR.

Horizontal and vertical micro-instructions

The arrangement of the micro-instruction format above implies that all possible control signals are specified in parallel in each micro-instruction. Such an organisation of control bits permits the highest possible degree of concurrency to be achieved: a number of simultaneous operations can be specified within one micro-instruction. This format is called a *horizontal* micro-instruction. It extracts the fastest possible speed of operation for a given architecture and requires no decoding of the control outputs. However, since in practice there are very many control lines — of the order of a hundred or more — the horizontal format leads to very long control memory words which can be impracticable to implement.

At the opposite extreme is a method of encoding control signals within the micro-instruction so that only certain combinations of control outputs can be specified at one time. This reduces the degree of parallelism which can be achieved by one micro-instruction, thus making necessary several micro-instructions to implement a required action using this method, compared to only one using the horizontal format. The width of the control field, and thus of the micro-instruction itself, can be minimised, giving a more manageable control memory word length. However, decoding circuits have to be used to convert the control field contents of the CMDR to suitable control signals. This type of format, the *vertical* micro-instruction, leads to slower speeds of operation because of the lack of parallelism and the need to decode control outputs. It may, on the other hand, be much more efficient in its use of control memory space because very often a relatively few control signal combinations require to be specified together and in such cases the horizontal format uses unnecessarily large amounts of memory.

In practice a scheme somewhere between the two extremes is normally used, and will be chosen to give a compromise between the parallelism (higher speed) of the horizontal format and the control field encoding (smaller control memory) of the vertical arrangement.

CHAPTER 6

The hardware/software interface

6.1 INTERDEPENDENCE OF HARDWARE AND SOFTWARE

From the programming point of view, computer hardware is represented by a set of machine-code instructions. The structure of the underlying hardware is otherwise evident in the number and nature of user-programmable registers which are an integral part of the instruction set.

Indeed, it is much more common for the machine-code instruction level to be completely hidden: most programming languages are *high-level,* specifically aimed at removing machine-dependent features from the programmer's province, and most users of a computer see it at a higher level still — as indicated by the levels in Fig.1.1. Programmers and users are mainly concerned with solving problems and using computers to help them in their everyday tasks, and the details of the machine architecture are of no concern to them. Nevertheless, hardware and software are closely interdependent: software and general user requirements dictate the hardware features, and implicitly the logical structure, of computers; on the other hand their technology and architecture have a very strong bearing on the form and efficiency of the software.

In terms of layers, hardware and software meet at the machine-code instruction level in a hardwired computer, but at the microprogramming level in the more flexible microprogrammed type of machine. Since machine-code instruction sets can be altered easily by microprogramming, it is possible to *emulate* the instruction set of different computers on a microprogrammed computer, effectively giving the programmer (at the machine-code level) a different machine to program. Emulation is not simply a matter of changing the machine-code instruction set but it also depends heavily on the compatibility or otherwise of the registers and the word lengths of the *host* (the machine undergoing the microprogram alterations) and the *target* (the computer being emulated). Emulation permits programs written for one computer to be run on another — perhaps with a loss of efficiency if the host and target are incompatible — and so allows the ideal of *software portability*. Users with a heavy investment in computer programs cannot afford to have to rewrite software on a large scale when upgrading or changing hardware, therefore it is highly desirable to be able to transport all the existing software from the old to the new computer. The availability of microprogrammed machines with an effectively flexible architecture is thus one way in which the structure of computer hardware is influenced by software requirements, in this case by the aim of portability.

Computers are, as we have seen, *multi-level* machines. At the lowest level is the hardware, consisting of the logic circuits in the CPU, memory and input/output units. Hardware and software meet at either the microprogramming or machine-code levels, which are intermediate in level between the hardware and the high-level programming language. Machine code is used for writing operating system software, at least the machine-dependent parts of it. Applications software, and the upper layers of the operating system, tend to be written in high-level languages. A multi-level machine is illustrated in Fig. 6.1, including the uppermost (user) level. Proceeding from the lowest level upwards, each offers a different *virtual machine* interface to the next higher level: below the interface there appears to be a machine with characteristics different from the basic hardware. From the top. the user sees a set of task-oriented commands offered by the applications machine, which in turn has a view of a high-level language machine; supporting the high-level language interface is the underlying operating system, which is built on top of the machine-code instruction level. If the microprogramming level exists it forms a layer between the machine-code level and the basic hardware machine consisting of logic circuits and gates. Otherwise the machine-code level maps directly onto the hardware.

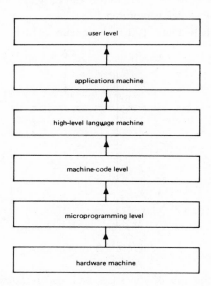

Fig. 6.1 – A multi-level machine.

In practice the characteristics of any level may affect not only the next higher one but all those above. In particular the form of the computer architecture — consisting of the hardware machine and its microprogramming level — has an influence on the software and user levels, from the machine-code right up to the user level. The micro-instruction set and machine registers dictate the possible machine-code instructions, which in turn have an effect on the efficiency, or even the implementability, of high-level languages. It is the efficiency (in running time and required memory size) of programs written in high-level languages which determines the suitability of a computer for an application. Choice of computer to do a particular job usually depends on the total machine from the hardware to the high-level machine inclusive.

Conversely, the requirements of each of the software levels, from user down to machine-code, have an influence on the underlying computer architecture. Over the generations of computers, a variety of architectural features has emerged, all a result of higher-level demands. We shall explore some of these to illustrate the dependence of computer design on the software and user levels.

The influence of high-level languages on computer design is evident from a number of factors. The most often quoted of these is the use of the *stack*. In high-level languages arithmetic expressions consist of a series of operands and operators in infix form, for example the product of A and B is written as A * B, that is the multiplication operator is situated in between the two operands. Arbitrarily long expressions may usually be written in a high-level programming statement, and these must be evaluated as the program runs. To evaluate infix expressions the use of a stack is required.

A stack is essentially a last-in, first-out queue, as illustrated in Fig. 6.2.

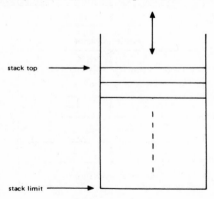

stack top

stack limit

Fig. 6.2 — Outline of a stack.

Items are added at the top of the stack and also removed from the top. The stack grows and diminishes as items are added and removed; it is fixed at one end, called the stack limit. A stack can easily be implemented in a computer memory by setting the stack limit at a fixed address, and allowing the stack to grow away from this address. A stack pointer register is usually implemented in hardware to contain the address of the current top of the stack. Machine-code instructions are provided which automatically adjust the value of the stack pointer when items are added or removed. Computers with such a stack architecture have been built specifically to help the efficient implementation of high-level programming languages.

Stacks are also commonly employed in computers to implement *subroutine* call and return statements. Subroutines are a feature at the machine-code level of the majority of modern computers, and allow the programming of a large task to be split up into a number of smaller program modules. Generally there would be a main program module and a number of subroutines, each called from the main program. When the subroutine call is obeyed, the value of the PC register, which will be pointing to the instruction after the call, is put onto the stack and the PC register loaded with the address of the subroutine. At the end of the subroutine will be a return instruction; when this is obeyed, it has the effect of removing the address which had been stored on the stack earlier in the process. The address is loaded into the PC, and the main program resumed. Subroutines within subroutines may be implemented since the use of the stack ensures that return addresses are stored in the correct order.

Another use of stacks is in connection with the *interrupt* mechanism which most computers possess. Although not stated in earlier chapters, usually the CPU is capable of being interrupted by external events, in order to implement *multiprogramming*. It is extremely important that the resources of the CPU are used as efficiently as possible, and this can be attained by constantly ensuring that the highest priority program runs on the CPU. Originally multiprogramming was introduced to optimise the throughput of work by changing the running program when it requested an input/output operation. I/O generally takes a very long time compared to the time required to obey a machine-code instruction, so a program waiting for completion of I/O would waste CPU time unless replaced by another. Thus multiprogramming ensures full utilisation of processor time. When a program runs on a CPU its *context* consists of the values stored in the machine registers. Removing a program from the CPU involves temporary storage of the register values, which must be subsequently replaced in the registers when that program is reloaded. This storing and reloading may be effected by a succession of stack operations. In modern computers the completion of I/O operations is an external event which causes an interrupt to the processor, and an automatic stacking of the context of the running program. The interrupted program is replaced by an interrupt service routine which determines the cause of the interrupt, and deals with it appropriately. Generally this routine then calls the operating system *scheduler* which may either reload the interrupted program, or load another having higher priority. Other sources of interrupts include timer and fault events: a running program is interrupted if it has run for more than a maximum time-slot or if the computer hardware detects a possible fault condition.

Information is represented in high-level language programming in one of a number of forms: as integers, or real numbers, or characters, in simple cases. Other cases may include Booleans and arrays. The programmer distinguishes between objects not only by name, but also by *type*. Generally speaking operations on mixed types are not permitted. The main reason for having types in languages apart from improving their readability, is to ensure software *reliability*. A programmer is less likely to write incorrect programs where types are used, and likewise it is easy for the *compiler* (the program which translates high-level language into machine-code) to perform type-checks at translation time. However, the likelihood of errors is reduced further if a *tagged* architecture is used. This involves the labelling of each item of information with a tag to indicate its type: this tag is carried about with the value of the item. Thus at run time the hardware can also perform type-checks. Although a tagged architecture may seem a desirable feature, surprisingly few computers have been built in this way.

High-level programming languages are intended to be *problem-oriented*, and ideally they should disguise the characteristics of the underlying hardware. One of the chief limitations of any computer is its addressing capability, that is the amount of memory which can be addressed by the CPU. The limit is set by the word length of the computer, as explained in Chapter 1. Limiting the amount of memory means that programs of only a certain maximum size can be stored. In a multi-programming machine many programs will concurrently share the main memory, and the amount of space each can occupy can be severely restricted. When a programmer writes in high-level language, nothing should be known of the addressing limitations. Yet the problem exists: the computer must be able to cope with large and small programs alike, invisibly to the programmer. A scheme whereby the main memory and the (much larger) backing store — magnetic discs and tape — may appear to be combined into one substantial memory is implemented on many computers. It is called *virtual memory* (VM). The original idea of VM was developed at Manchester University when the ATLAS computer was designed with a *1-level store,* in the early 1960s. It was a decade later when commercial machines incorporating VM became more common. The VM scheme is illustrated in Fig. 6.3.

Addresses generated by the programmer are said to be in *name* space (numerical addresses will be produced by the compiler). These are mapped (as shown above) automatically by the hardware into *memory* space, physically consisting of the more immediately accessible main memory and the slower backing store. Such a mapping mechanism is incorporated in the design of the computer. The usual form of implementation of VM is called *paging,* in which main memory, and backing store, are divided into a number of fixed size areas called pages. A program will consist of one or more pages, not all of which are necessarily in the main memory at the same time. The entire program may be scattered over regions of main memory and backing store. It is the VM management mechanism which keeps track of all the pages belonging to the various programs. The address translation is normally implemented using an *associative memory* which contains a list of all the pages currently in main memory, and their corresponding physical addresses. Pages on backing stores are not represented in this associative memory. When an address is generated within a program at run-time part of the address, the page number, is presented to the associative memory. If the page is present, its physical page address is produced and summed with the word number from the original address. The final effective address is passed to the MAR register, and used to fetch the required word from main memory. This scheme is illustrated in Fig. 6.4.

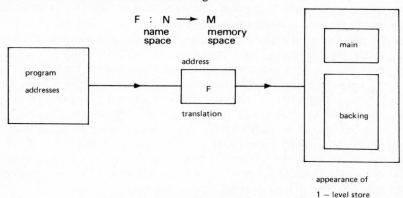

Fig. 6.3 — Outline of virtual memory scheme.

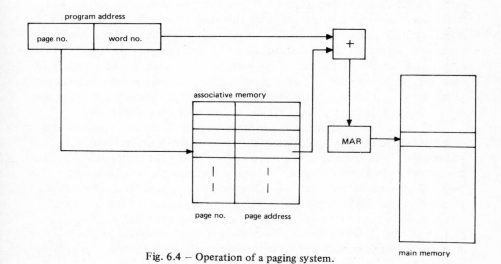

Fig. 6.4 — Operation of a paging system.

If the page number is not present in the associative memory a page fault interrupt is generated and a service routine arranges to fetch the required page from backing store. Obviously this scheme requires a more complex memory management system than the simple MAR, MDR arrangement previously illustrated (Chapter 1). Of central importance is the associative memory itself. This is otherwise called a *content-addressable* memory, which better summarises its operation. Operands are not accessed in the memory by their relative address from the start of the memory but by their contents. Access times using this system are very fast, although the cost of associative memory hardware is high enough to restrict its widespread adoption.

Use of a VM system can help improve the work throughput of a computer since, in the von Neumann machine, memory accesses can cause a bottleneck in its operation, and an automatic memory management system can be tuned to minimise the number of backing store accesses. The *working set model* attributed to P. J. Denning is a strategy used in conjunction with VM which aims to allow each program to keep a minimum number of pages in main memory at once, by monitoring and controlling the pages accessed in a given time interval. Only those pages accessed within this time interval are allowed to remain in memory, while the others are transferred to backing store. This avoids the possibility that too many programs, each with too few pages for their needs, will occupy the memory concurrently: such a state of affairs causes the problem of *thrashing* whereby almost all of the computer's time is spent transferring pages between main memory and backing store. The working set model, too, is a neglected architectural feature despite its apparent attractiveness.

Cache memories are also used to minimise accesses to main memory, and can speed up the operation of a computer typically two or three times. The cache device is situated between the CPU and the main memory, and is a smaller, faster-access memory intended to store the most frequently accessed items of information. A characteristic feature of program operation is that a small percentage of items tend to be accessed for a high percentage of the time. A cache only a few percent of the size of the main memory is capable of a *hit rate* of over 90%, in other words on average 90% of requests to the main memory are satisfied by information contained in the cache. Items found in the cache are (very rapidly) transferred to the CPU and the main memory undisturbed.

The general user requirements for higher speeds of operation also influence computer architecture. Apart from the use of ever-faster integrated circuit technologies, computers can be speeded up by a number of techniques including the cache memory just described. The use of *parallelism* and *pipelining* has been reported for some larger computers, although not in general for smaller, cheaper machines: these architectural features tend to be expensive in hardware. Parallelism effectively aims to use multiple processing elements instead of a single CPU, to take advantage of the inherent parallelism of many computational problems. A typical example of such a problem is matrix arithmetic, which occurs in many computer applications. The idea is that each of the processing elements is given a part of the problem to evaluate, all the operations being carried out in parallel. Each processing element usually has its own memory. Fast processors based on this technique are available commercially for applications which require the speed and can justify the cost. Pipelining, on the other hand, attempts to take advantage of the fact that a machine-code instruction consists of several smaller steps (microinstructions) which can often be overlapped in time. In particular the next instruction can be fetched while the present instruction is being obeyed. Since memory fetches take a relatively long time, pipelining — which basically overlaps fetch and obey steps — can significantly improve the running speed of programs. Pipelining is commonly used in the larger mainframes.

Less obviously, the use of *optimised instruction sets* can be beneficial to the speed of operation of a program. Just as in memory accessing a few items are accessed most of the time, so in practice a relatively few machine-code instructions are used a large percentage of the time. It makes sense to optimise the performance of these machine-code instructions at the expense of all the others: the microprogramming level can be arranged to meet this requirement. User-microprogrammable computers are particularly well suited to producing more efficient operation using this technique: for a particular application it can be found which instructions are most frequently used, and these can be appropriately re-microprogrammed.

6.2 SUMMARY

Despite the various architectural features described in the previous section, computers still share the basic von Neumann structure introduced in Chapter 1. The requirements of higher speed and reliability, and the demands of programming languages, have not dramatically altered the way in which program instructions are fetched and obeyed. Nevertheless it might be appropriate to conclude this introduction to the subject of computer structure by speculating briefly on likely advances in design and construction in the immediate future.

One general point concerns the relationship of technology to computer architecture. Over the years since integrated circuits first appeared the number of components on a chip has increased very rapidly. At the same time speeds have increased and costs decreased. These advances in technology mean that more and more powerful CPUs and larger and larger memories can be fabricated at minimal cost on a single chip. Development of this kind have significantly increased the number of application areas in which computers are used, most notably those in industrial and domestic control, and in entertainment. However they have not of themselves brought advances in computer architecture.

On the other hand, LSI and VLSI technologies have had a notable impact on the logic design techniques of computer systems. Large-scale design has become the province of the semiconductor manufacturer: complex devices such as microprocessors may still be conceived in terms of gates and flip-flops, but are implemented by implanting layers on a silicon chip. The traditional minimisation aims no longer apply, but are replaced by other criteria such as component packing density, the size of silicon chips which can be made without flaws and the requirements for encapsulating the product in an integrated circuit package. The application designer may now think in terms of much higher level building blocks, and in some cases the way they are connected to implement the application can be specified without resorting to traditional logic design techniques.

Advances in IC technology will initially continue at the same pace as before. In the foreseeable future the trends will continue: larger packing densities will be achieved, and the rate of increase of operating speeds may be maintained (although this rate will decrease as the limiting factor of the speed of light is realised), but the problem of chip reliability will inevitably grow.

Advances in computer architecture are likely to be the result of the requirements of new forms of programming language. The present high-level languages, although independent of the features of a particular machine hardware, nevertheless reflect the conventional von Neumann computer structure. Until language designers can produce a radically different form for expressing programming requirements, it is difficult to see equally radical change in hardware design taking place. This change, when it comes, will be in response to the need to narrow the gap between software and hardware.

Reading list

The following list gives recommendations for further reading on a Chapter-by-Chapter basis.

Chapter 1

Meek, B. L. and Fairthorne, S. *Using computers,* Ellis Horwood, 1977.
> An introduction to the nature and applications of computers. Provides suitable background material in hardware and software.

Colin, A. J. T. *Fundamentals of computer science,* Macmillan, 1980.
> An alternative introduction to computers and their structure.

Bowden, B. V. (Ed.) *Faster than thought,* Pitman, 1953; paperback edition, 1971.
> A series of papers published in the early days of computing, on the history, structure and applications of computers.

Randell, B. (Ed.) *The origins of digital computers: selected papers,* Springer Verlag, 1973.
> Traces the history of computers from before Babbage up to the first generation of stored-program electronic machines. The story is told by a set of carefully selected papers, each reporting significant steps in computer developments, and each prefaced by a commentary by the editor.

Burks, A. W., Goldstine, H. H. and von Neumann, J. *Preliminary discussion of the logical design of an electronic computing instrument,* U.S. Army Ordnance Report, 1946; reprinted in Bell, C. G. and Newell, A. *Computer structures: readings and examples,* McGraw-Hill, 1971.
> This two-part report describes the features of the so-called von Neumann machine, a description which still applies to today's computers.

Bardeen, J. and Brattain, W. H. *The transistor, a semiconductor triode, Phys. Rev.,* vol. 74, pp. 230–231, 1948.
> Report of the development of the transistor.

Denning, P. J. *Third generation computer systems, A.C.M. Computing Surveys,*
vol. 3, pp. 175–216, 1971.
> Excellent review of computers of the third generation, describing both
> hardware and software developments of the previous two generations.
Langdon, G. G. *Logic design: a review of theory and practice,* Academic Press,
1974.
> Useful historical review of developments in logic design applied to
> computers. Contains material relevant also to Chapter 5.

Chapter 2

Microelectronics, Scientific American, September 1977; republished by W. H.
Freeman, 1978.
> An issue devoted to the structure and applications of microelectronic
> circuits — how integrated circuits, particularly microprocessors and
> memories, are fabricated and incorporated in computer systems.
Mazda, F. F. *Integrated circuits: technology and applications,* Cambridge Uni-
versity Press, 1978.
> A short description of integrated circuit fabrication, logic families and
> the use of integrated circuits in practical applications.
Morris, R. L. and Miller, J. R. (Eds.) *Designing with TTL integrated circuits,*
McGraw-Hill, 1971.
> Written by the staff of Texas Instruments, this book is particularly
> valuable for its description of the TTL family. The design examples
> also provide useful material for Chapter 4.
The TTL data book for design engineers, Texas Instruments Limited, 2nd
edition, 1977.
> Complete selection guide for TTL integrated circuits. Extracts from this
> publication are presented in the Appendix.
The interface circuits data book, Texas Instruments Limited, 1978.
> Complete selection guide for interface circuits available from Texas
> Instruments. See also *IC Master 1979.*
The memory and microprocessor data book, Texas Instruments Limited, 1976.
> A third selection guide from Texas Instruments, more specialised than
> the other two since it describes only a narrow range of (Texas) micro-
> processors and memories.
IC Master 1979, United Technical Publications Inc., 1979.
> A publication from the U.S.A. containing a full list of available inte-
> grated circuits, produced independently of any semiconductor manu-
> facturer. The broad IC categories are digital, linear, interface, memory
> and microprocessor.
Boole, G. *An investigation of the laws of thought,* Macmillan, 1854; reprinted
by Dover, 1958.
> Boole's work is the foundation of the modern techniques of logic

design, although the form in which its results are expressed is unpalatable to the modern reader.

Huntington, E. V. *Sets of independent postulates for the algebra of logic, Trans. American Mathematical Society,* vol. 5, pp. 288–309, 1904.

This paper re-expresses the basic rules of Boolean algebra in a more modern form.

Shannon, C. E. *A symbolic analysis of relay and switching circuits, Trans. AIEE,* vol. 57, pp. 713–723, 1938.

In this paper Shannon shows how relay and switching circuits can be described consistently with the rules of Boolean algebra. The techniques presented are used in modern digital logic design.

Chapter 3

Hill, F. J. and Peterson, G. R. *Introduction to switching theory and logical design,* Wiley, 2nd edition, 1974.

A well-established text on logic design. Gives a full account of the theory of design techniques.

Lewin, D. *Logical design of switching circuits,* Nelson, 2nd edition, 1974.

Recommended for its sections on synchronous and asynchronous sequential circuit design, and on logic design using MSI components. Contains material also to supplement Chapter 4.

Mowle, F. J. *A systematic approach to digital logic design,* Addison-Wesley, 1976.

Contains an excellent description of the theory of Boolean algebra, and has many useful logic design examples throughout the text.

Karnaugh, M. *The map method for synthesis of combinational logic circuits, Trans. AIEE,* vol. 72, pp. 593–599, 1953.

Original description of the K-map method for representing (and synthesising) combinational circuits.

Quine, W. V. *The problem of simplifying truth functions, Am. Math. Mon.,* vol. 59, pp. 521–531, 1952.

This paper presents the original ideas for the tabular method of minimisation of logic functions.

McCluskey, E. *Minimization of Boolean functions, Bell Syst. Tech. Journal,* vol. 35, pp. 1417–1444, 1956.

Developing Quine's ideas, describes the tabular minimisation method now known as the Quine-McCluskey method.

Moore, E. F. *Gedanken experiments on sequential machines, Automation Studies (Ann. Math. Studies* No. 34), Princeton University Press, 1956.

Describes a theoretical model of a sequential logic circuit which has proved extremely useful to designers.

Mealy, G. H. *A method of synthesizing sequential circuits, Bell Syst. Tech. Journal*, vol. 34, pp. 1045-1079, 1955.
> Describes an alternative model of a sequential logic circuit. The ideas in this paper and Moore's are frequently combined into the collective Moore/Mealy model.

Minsky, M. *Computation: finite and infinite machines*, Prentice-Hall, 1972.
> An excellent text on the theory of computation. Recommended for further reading on finite-state machines.

Chapter 4

Zissos, D. *Problems and solutions in logic design*, Oxford University Press, 1976.
> An excellent collection of problems with worked solutions in both combinational and sequential logic.

Lewin, D. *Logical design of switching circuits*, Nelson, 2nd edition, 1974.
> Contains useful additional reading on circuit hazards.

Morris, R. L. and Miller, J. R. (Eds.) *Designing with TTL integrated circuits*, McGraw-Hill, 1971.
> Contains a full explanation of the fanout capabilities of TTL circuits.

Tocci, R. *Digital systems: principles and applications*, Prentice-Hall, revised edition, 1980.
> Contains useful and well-illustrated sections on circuit problems in practice.

Peatman, J. B. *Microcomputer-based design*, McGraw-Hill, 1977.
> Despite the title, includes valuable sections on the practical uses of integrated circuits. The use of microprocessors (as universal logic elements) in control applications makes interesting, and relevant, reading.

Chapter 5

Hayes, J. P. *Computer architecture and organization*, McGraw-Hill, ISE, 1978.
> Contains an excellent description of computer design methodology. The uses of logic circuits are well illustrated. Recommended also for its full section on the overall organisation of computer systems.

Boulaye, G. G. *Microprogramming*, Macmillan, 1975.
> A good brief account of the principles and techniques of microprogramming.

Langdon, G. G. *Logic design: a review of theory and practice*, Academic Press, 1974.
> Gives historical account of the techniques of computer design. Of particular relevance is the discussion of the use of timing signals within computers.

Bell, C. G. and Newell, A. *Computer structures: readings and examples,* McGraw-Hill, 1971.

> Contains a wealth of information about the detailed structure of various different computers. Shows the variety of computer architectures which have been devised.

Tanenbaum, A. S. *Structured computer organization,* Prentice-Hall, 1976.

> The notion of levels within a computer system is strongly emphasised. The microprogramming level is particularly well described.

Wilkes, M. V. *The best way to design an automatic calculating machine, Rept. Manchester University Computer Inaugural Conf.,* pp. 16–18, 1951; reprinted in Swartzlander, E. E. (Ed.) *Computer design development: principal papers,* pp. 266–270, Hayden, 1976.

> The original paper on microprogramming.

Chapter 6

Donovan, J. J. *Systems programming,* McGraw-Hill, ISE, 1972.

> A fairly complete text on system software. Particularly recommended are the sections on machine language and assemblers. Relates software and computer architecture clearly, despite being based on a specific mainframe.

Lister, A. M. *Fundamentals of operating systems,* Macmillan, 2nd edition, 1980.

> The best available brief account of operating systems structure.

Brown, P. J. *Macro processors and techniques for portable software,* Wiley, 1974.

> A good introduction to macros and more specifically to their use in implementing virtual machine architectures.

Tanenbaum, A. S. *Structured computer organization,* Prentice-Hall, 1976.

> The account of multi-level machines is strongly recommended.

Wilkes, M. V. *Time-sharing computer systems,* Macdonald and Jane's, 3rd edition, 1975.

> The subjects of virtual memory and protection are briefly, and well, described.

Denning, P. J. *Fault tolerant operating systems, A.C.M. Computing Surveys,* vol. 8, pp. 359–389, 1976.

> A review paper which describes the implications for computer architecture of implementing fault-tolerant systems.

Anderson, T. and Randell, B. (Eds.) *Computing systems reliability,* Cambridge University Press, 1978.

> A collection of papers covering all aspects of hardware and software reliability.

Davies, D. W. and Barber, D. *Communications networks for computers,* Wiley, 1973.

 Highly recommended descriptions of computer network architectures.

Myers, G. T. *Advances in computer architecture,* McGraw-Hill, 1978.

 Highlights the need for new architectures which are more suitable for implementing software requirements.

Denning, P. J. *Why not innovations in computer architecture?, A.C.M. Computer Architecture News,* vol. 8, pp. 4–7, 1980.

 A short discussion of the reasons why conventional computer architectures predominate. Includes useful further references.

Appendix

The following pages are reprinted, by kind permission of Texas Instruments Limited[†], from *The TTL data book for design engineers* (2nd edition). The material is representative of digital semiconductor product information and comprises:

(1) an index to selection guide information for SSI and MSI/LSI functions;

(2) five sample selection guide pages, two for SSI and three for MSI/LSI products;

(3) two pages from the TTL integrated circuits mechanical data section, the first giving general ordering information, the second showing specifications of a 14-pin plastic dual-in-line integrated circuit package;

(4) five sample data sheet pages showing the pin configurations and functions of a number of integrated circuits – these pages are each referenced in one of the selection guide pages in (2) above.

Texas Instruments Limited is a semiconductor *manufacturer*. Their products and literature, in common with those of other manufacturers, are available mainly through a number of appointed *distributors*, such as Quarndon Electronics (Semiconductors) Limited of Derby, or Celdis of Reading.

†Texas Instruments Limited,
Northern European Semiconductor Division
Manton Lane
Bedford MK41 7PA
England

FUNCTIONAL INDEX/SELECTION GUIDE

The following pages contain functional indexes and selection guides designed to simplify the choice of a particular function to fit a specific application. Essential characteristics of similar or like functions are grouped for comparative analysis, and the electrical specifications are referenced by page number. The following categories of functions are covered:

SSI FUNCTIONS

Page

Positive-NAND gates and inverters with totem-pole outputs 1-10
Positive-NAND gates and inverters with open-collector outputs 1-10
Positive-NOR gates with totem-pole outputs 1-11
Positive-AND gates with totem-pole outputs 1-11
Positive-AND gates with open-collector outputs 1-11
Schmitt-trigger positive-NAND gates and inverters with totem-pole outputs 1-11
Buffers/clock drivers with totem-pole outputs 1-12
50-ohm/75-ohm line drivers 1-12
Buffer and interface gates with open-collector outputs 1-12
Gates, buffers, drivers and bus transceivers with 3-state outputs 1-13
Positive-OR gates with totem-pole outputs 1-14
AND-OR-INVERT gates with totem-pole outputs 1-14
AND-OR-INVERT gates with open-collector outputs 1-14
Expandable gates 1-14
Expanders 1-14
Dual J-K edge-triggered flip-flops 1-15
Single J-K edge-triggered flip-flops 1-15
Pulse-triggered dual flip-flops 1-16
Pulse-triggered single flip-flops 1-16
Dual J-K flip-flops with data lockout 1-16
Single J-K flip-flops with data lockout 1-16
Dual D-type flip-flops 1-16
S-R latches 1-17
Current-sensing gates 1-17
Monostable multivibrators with Schmitt-trigger inputs 1-17
Retriggerable monostable multivibrators 1-17
Clock generator circuits 1-17

MSI/LSI FUNCTIONS

Adders 1-18
Accumulators, arithmetic logic units, look-ahead carry generators 1-18
Multipliers 1-18
Comparators 1-18
Parity generators/checkers 1-19
Other arithmetic operators 1-19
Quad, hex, and octal flip-flops 1-19
Register files 1-19
Shift registers 1-20
Other registers 1-20
Latches 1-21
Clock generator circuits 1-21
Code converters 1-21
Priority encoders/registers 1-22
Data selectors/multiplexers 1-22
Decoders/demultiplexers 1-23
Open-collector display decoders/drivers with counters/latches 1-23
Open-collector display decoders/drivers 1-24
Bus transceivers and drivers 1-25
Asynchronous counters (ripple clock)—negative-edge triggered 1-25
Synchronous counters—Positive-edge triggered 1-26
Bipolar bit-slice processor elements 1-26
First-in first-out memories (FIFO's) 1-26
Random-access read/write memories (RAM's) 1-27
Read-only memories (ROM's) 1-27
Programmable-read-only memories (PROM's) 1-28
Microprocessor controllers and support functions 1-28

TEXAS INSTRUMENTS

SSI FUNCTIONS
FUNCTIONAL INDEX/SELECTION GUIDE

POSITIVE-NAND GATES AND INVERTERS WITH TOTEM-POLE OUTPUTS
ELECTRICAL TABLES — PAGE 6-2

DESCRIPTION	TYPICAL PROPAGATION DELAY TIME	TYP POWER DISSIPATION PER GATE	DEVICE TYPE AND PACKAGE				PIN ASSIGNMENTS PAGE NO.
			−55°C to 125°C		0°C to 70°C		
HEX INVERTERS	3 ns	19 mW	SN54S04	J, W	SN74S04	J, N	
	6 ns	22 mW	SN54H04	J, W	SN74H04	J, N	
	9.5 ns	2 mW	SN54LS04	J, W	SN74LS04	J, N	5-7
	10 ns	10 mW	SN5404	J, W	SN7404	J, N	
	33 ns	1 mW	SN54L04	J, T	SN74L04	J, N	
QUADRUPLE 2-INPUT POSITIVE-NAND GATES	3 ns	19 mW	SN54S00	J, W	SN74S00	J, N	
	6 ns	22 mW	SN54H00	J, W	SN74H00	J, N	
	9.5 ns	2 mW	SN54LS00	J, W	SN74LS00	J, N	5-6
	10 ns	10 mW	SN5400	J, W	SN7400	J, N	
	33 ns	1 mW	SN54L00	J, T	SN74L00	J, N	
TRIPLE 3-INPUT POSITIVE-NAND GATES	3 ns	19 mW	SN54S10	J, W	SN74S10	J, N	
	6 ns	22 mW	SN54H10	J, W	SN74H10	J, N	
	9.5 ns	2 mW	SN54LS10	J, W	SN74LS10	J, N	
	10 ns	10 mW	SN5410	J, W	SN7410	J, N	
	33 ns	1 mW	SN54L10	J, T	SN74L10	J, N	
DUAL 4-INPUT POSITIVE-NAND GATES	3 ns	19 mW	SN54S20	J, W	SN74S20	J, N	
	6 ns	22 mW	SN54H20	J, W	SN74H20	J, N	
	9.5 ns	2 mW	SN54LS20	J, W	SN74LS20	J, N	5-10
	10 ns	10 mW	SN5420	J, W	SN7420	J, N	
	33 ns	1 mW	SN54L20	J, T	SN74L20	J, N	
8-INPUT POSITIVE-NAND GATES	3 ns	19 mW	SN54S30	J, W	SN74S30	J, N	
	6 ns	22 mW	SN54H30	J, W	SN74H30	J, N	
	17 ns	2.4 mW	SN54LS30	J, W	SN74LS30	J, N	5-12
	10 ns	10 mW	SN5430	J, W	SN7430	J, N	
	33 ns	1 mW	SN54L30	J, T	SN74L30	J, N	
13-INPUT POSITIVE-NAND GATES	3 ns	19 mW	SN54S133	J, W	SN74S133	J, N	5-38

POSITIVE-NAND GATES AND INVERTERS WITH OPEN-COLLECTOR OUTPUTS
ELECTRICAL TABLES — PAGE 6-4

DESCRIPTION	TYPICAL PROPAGATION DELAY TIME	TYP POWER DISSIPATION PER GATE	DEVICE TYPE AND PACKAGE				PIN ASSIGNMENTS PAGE NO.
			−55°C to 125°C		0°C to 70°C		
HEX INVERTERS	5 ns	17.5 mW	SN54S05	J, W	SN74S05	J, N	
	8 ns	22 mW	SN54H05	J, W	SN74H05	J, N	5-7
	16 ns	2 mW	SN54LS05	J, W	SN74LS05	J, N	
	24 ns	10 mW	SN5405	J, W	SN7405	J, N	
QUADRUPLE 2-INPUT POSITIVE-NAND GATES	5 ns	17.5 mW	SN54S03	J, W	SN74S03	J, N	5-7
	8 ns	22 mW	SN54H01	J, W	SN74H01	J, N	5-6
	16 ns	2 mW	SN54LS01	J, W	SN74LS01	J, N	5-6
	16 ns	2 mW	SN54LS03	J, W	SN74LS03	J, N	5-7
	22 ns	10 mW	SN5401	J, W	SN7401	J, N	5-6
	22 ns	10 mW	SN5403	J	SN7403	J, N	5-7
	46 ns	1 mW	SN54L01	T			5-6
	46 ns	1 mW	SN54L03	J	SN74L03	J, N	5-7
TRIPLE 3-INPUT POSITIVE-NAND GATES	16 ns	2 mW	SN54LS12	J, W	SN74LS12	J, N	5-9
	22 ns	10 mW	SN5412	J, W	SN7412	J, N	
DUAL 4-INPUT POSITIVE-NAND GATES	5 ns	17.5 mW	SN54S22	J, W	SN74S22	J, N	
	8 ns	22 mW	SN54H22	J, W	SN74H22	J, N	5-11
	16 ns	2 mW	SN54LS22	J, W	SN74LS22	J, N	
	22 ns	10 mW	SN5422	J, W	SN7422	J, N	

TEXAS INSTRUMENTS

SSI FUNCTIONS
FUNCTIONAL INDEX/SELECTION GUIDE

PULSE-TRIGGERED DUAL FLIP-FLOPS

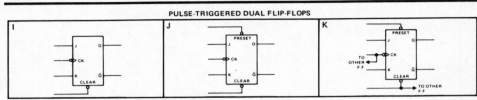

PULSE-TRIGGERED SINGLE FLIP-FLOPS

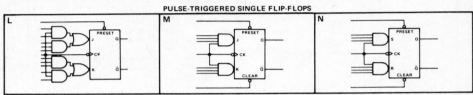

DWG. REF.	TYPICAL CHARACTERISTICS		DATA TIMES		DEVICE TYPE AND PACKAGE				PAGE REFERENCES	
	f_{max} (MHz)	Pwr/F-F (mW)	SETUP (ns)	HOLD (ns)	−55°C to 125°C		0°C to 70°C		PIN ASSIGNMENTS	ELECTRICAL
I	30	80	0↑	0↓	SN54H73	J, W	SN74H73	J, N	5-22	6-50
	20	50	0↑	0↓	SN5473	J, W	SN7473	J, N	5-22	6-46
	20.	50	0↑	0↓	SN54107	J	SN74107	J, N	5-32	6-46
	3	3.8	0↑	0↓	SN54L73	J, T	SN74L73	J, N	5-22	6-54
J	30	80	0↑	0↓	SN54H76	J, W	SN74H76	J, N	5-23	6-50
	20	50	0↑	0↓	SN5476	J, W	SN7476	J, N	5-23	6-46
K	30	80	0↑	0↓	SN54H78	J, W	SN74H78	J, N	5-24	6-50
	3	3.8	0↑	0↓	SN54L78	J, T	SN74L78	J, N	5-24	6-54
L	30	80	0↑	0↓	SN54H71	J, W	SN74H71	J, N	5-21	6-50
M	30	80	0↑	0↓	SN54H72	J, W	SN74H72	J, N	5-22	6-50
	20	50	0↑	0↓	SN5472	J, W	SN7472	J, N	5-22	6-46
	3	3.8	0↑	0↓	SN54L72	J, T	SN74L72	J, N	5-22	6-54
N	3	3.8	0↑	0↓	SN54L71	J, T	SN74L71	J, N	5-21	6-54

J-K FLIP-FLOPS WITH DATA LOCKOUT
DUAL   SINGLE

D-TYPE FLIP-FLOPS
DUAL

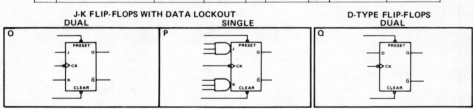

DWG. REF.	TYPICAL CHARACTERISTICS		DATA TIMES		DEVICE TYPE AND PACKAGE				PAGE REFERENCES	
	f_{max} (MHz)	Pwr/F-F (mW)	SETUP (ns)	HOLD (ns)	−55°C to 125°C		0°C to 70°C		PIN ASSIGNMENTS	ELECTRICAL
O	25	70	0↑	30↑	SN54111	J, W	SN74111	J, N	5-33	6-46
P	25	100	20↑	5↑	SN54110	J, W	SN74110	J, N	5-33	6-46
Q	110	75	3↑	2↑	SN54S74	J, W	SN74S74	J, N	5-22	6-58
	43	75	15↑	5↑	SN54H74	J, W	SN74H74	J, N	5-22	6-50
	33	10	25↑	5↑	SN54LS74A	J, W	SN74LS74A	J, N	5-22	6-56
	25	43	20↑	5↑	SN5474	J, W	SN7474	J, N	5-22	6-46
	3	4	50↑	15↑	SN54L74	J, T	SN74L74	J, N	5-22	6-54

↑↓The arrow indicates the edge of the clock pulse used for reference; ↑ for the rising edge, ↓ for the falling edge.

MSI/LSI FUNCTIONS
FUNCTIONAL INDEX/SELECTION GUIDE

ADDERS

DESCRIPTION	TYPICAL CARRY TIME	TYPICAL ADD TIME	TYP POWER DISSIPATION PER BIT	DEVICE TYPE AND PACKAGE −55°C to 125°C		0°C to 70°C		PAGE NO.
SINGLE 1-BIT GATED FULL ADDERS	10.5 ns	52 ns	105 mW	SN5480	J, W	SN7480	J, N	7-41
SINGLE 2-BIT FULL ADDERS	14.5 ns	25 ns	87 mW	SN5482	J, W	SN7482	J, N	7-49
SINGLE 4-BIT FULL ADDERS	10 ns	15 ns	24 mW	SN54LS83A	J, W	SN74LS83A	J, N	7-53
	10 ns	15 ns	24 mW	SN54LS283	J, W	SN74LS283	J, N	7-415
	11 ns	7 ns	124 mW	SN54S283	J	SN74S283	J, N	7-415
	10 ns	16 ns	76 mW	SN5483A	J, W	SN7483A	J, N	7-53
	10 ns	16 ns	76 mW	SN54283	J, W	SN74283	J, N	7-415
DUAL 1-BIT CARRY-SAVE FULL ADDERS	11 ns	11 ns	110 mW	SN54H183	J, W	SN74H183	J, N	7-287
	15 ns	15 ns	23 mW	SN54LS183*	J, W	SN74LS183*	J, N	7-287

ACCUMULATORS, ARITHMETIC LOGIC UNITS, LOOK-AHEAD CARRY GENERATORS

DESCRIPTION	TYPICAL CARRY TIME	TYPICAL ADD TIME	TYP TOTAL POWER DISSIPATION	DEVICE TYPE AND PACKAGE −55°C to 125°C		0°C to 70°C		PAGE NO.
4-BIT PARALLEL BINARY ACCUMULATORS	10 ns	20 ns	720 mW	SN54S281	J, W	SN74S281	J, N	7-410
4-BIT ARITHMETIC LOGIC UNITS/ FUNCTION GENERATORS	11 ns	20 ns	525 mW			SN74S381	N	7-484
	7 ns	11 ns	600 mW	SN54S181	J, W	SN74S181	J, N	7-271
	12.5 ns	24 ns	455 mW	SN54181	J, W	SN74181	J, N	7-271
	16 ns	24 ns	102 mW	SN54LS181	J, W	SN74LS181	J, N	7-271
LOOK-AHEAD CARRY GENERATORS	7 ns		260 mW	SN54S182	J, W	SN74S182	J, N	7-282
	13 ns		180 mW	SN54182	J, W	SN74182	J, N	

MULTIPLIERS

DESCRIPTION	DEVICE TYPE AND PACKAGE −55°C to 125°C		0°C to 70°C		PAGE NO.
2-BIT-BY-4-BIT PARALLEL BINARY MULTIPLIERS	SN54LS261	J, W	SN74LS261	J, N	7-380
4-BIT-BY-4-BIT PARALLEL BINARY MULTIPLIERS	SN54284, SN54285	J, W	SN74284, SN74285	J, N	7-420
	SN54S274	J	SN74S274	J, N	7-391
7-BIT-SLICE WALLACE TREES	SN54LS275	J	SN74LS275	J, N	7-391
	SN54S275	J	SN74S275	J, N	
25-MHz 6-BIT-BINARY RATE MULTIPLIERS	SN5497	J, W	SN7497	J, N	7-102
25-MHz DECADE RATE MULTIPLIERS	SN54167	J, W	SN74167	J, N	7-222

COMPARATORS

DESCRIPTION	TYPICAL COMPARE TIME	TYP TOTAL POWER DISSIPATION	DEVICE TYPE AND PACKAGE −55°C to 125°C		0°C to 70°C		PAGE NO.
4-BIT MAGNITUDE COMPARATORS	11.5 ns	365 mW	SN54S85	J, W	SN74S85	J, N	7-57
	21 ns	275 mW	SN5485	J, W	SN7485	J, N	
	23.5 ns	52 mW	SN54LS85	J, W	SN74LS85	J, N	
	82 ns	20 mW	SN54L85	J	SN74L85	J, N	

*New product in development as of October 1976.

TEXAS INSTRUMENTS

DECODERS/DEMULTIPLEXERS

DESCRIPTION	TYPE OF OUTPUT	TYPICAL SELECT TIME	TYPICAL ENABLE TIME	TYP TOTAL POWER DISSIPATION	DEVICE TYPE AND PACKAGE −55°C to 125°C		DEVICE TYPE AND PACKAGE 0°C to 70°C		PAGE NO.
4-LINE-TO-16-LINE	Totem-Pole	23 ns	19 ns	170 mW	SN54154	J, W	SN74154	J, N	7-171
	Totem-Pole	46 ns	38 ns	85 mW	SN54L154	J	SN74L154	J, N	7-171
	Open-Collector	24 ns	19 ns	170 mW	SN54159	J, W	SN74159	J, N	7-188
4-LINE-TO-10-LINE, BCD-TO-DECIMAL	Totem-Pole	17 ns		35 mW	SN54LS42	J, W	SN54LS42	J, N	
	Totem-Pole	17 ns		140 mW	SN5442A	J, W	SN7442A	J, N	7-15
	Totem-Pole	34 ns		70 mW	SN54L42	J	SN74L42	J, N	
4-LINE-TO-10-LINE, EXCESS-3-TO-DECIMAL	Totem-Pole	17 ns		140 mW	SN5443A	J, W	SN7443A	J, N	7-15
	Totem-Pole	34 ns		70 mW	SN54L43	J	SN74L43	J, N	
4-LINE-TO-10-LINE EXCESS-3-GRAY-TO-DECIMAL	Totem-Pole	17 ns		140 mW	SN5444A	J, W	SN7444A	J, N	7-15
	Totem-Pole	34 ns		70 mW	SN54L44	J	SN74L44	J, N	
3-LINE-TO-8-LINE	Totem-Pole	8 ns	7 ns	245 mW	SN54S138	J, W	SN74S138	J, N	7-134
	Totem-Pole	22 ns	21 ns	31 mW	SN54LS138	J, W	SN74LS138	J, N	7-134
DUAL 2-LINE-TO-4-LINE	Totem-Pole	7.5 ns	6 ns	300 mW	SN54S139	J, W	SN74S139	J, N	7-134
	Totem-Pole	22 ns	19 ns	34 mW	SN54LS139	J, W	SN74LS139	J, N	7-134
	Totem-Pole	18 ns	15 ns	30 mW	SN54LS155	J, W	SN74LS155	J, N	7-175
	Totem-Pole	21 ns	16 ns	125 mW	SN54155	J, W	SN74155	J, N	7-175
	Open-Collector	23 ns	18 ns	125 mW	SN54156	J, W	SN74156	J, N	7-175
	Open-Collector	33 ns	26 ns	31 mW	SN54LS156	J, W	SN74LS156	J, N	7-175

OPEN-COLLECTOR DISPLAY DECODERS/DRIVERS WITH COUNTERS/LATCHES

DESCRIPTION	OUTPUT SINK CURRENT	OFF-STATE OUTPUT VOLTAGE	TYP TOTAL POWER DISSIPATION	BLANKING	DEVICE TYPE AND PACKAGE −55°C to 125°C		DEVICE TYPE AND PACKAGE 0°C to 70°C		PAGE NO.
BCD COUNTER/ 4-BIT LATCH/ BCD-TO-DECIMAL DECODER/DRIVER	7 mA	55 V	340 mW				SN74142	J, N	7-140
BCD COUNTER/ 4-BIT LATCH/ BCD-TO-SEVEN-SEGMENT DECODER/ LED DRIVER	Constant Current 15 mA	7 V	280 mW	Ripple	SN54143	J, W	SN74143	J, N	7-143
BCD COUNTER/ 4-BIT LATCH/ BCD-TO-SEVEN-SEGMENT DECODER/ LAMP DRIVER	20 mA	15 V	280 mW	Ripple	SN54144	J, W			7-143
	25 mA	15 V	280 mW	Ripple			SN74144	J, N	

RESULTANT DISPLAYS USING '143, '144

0   1   2   3   4   5   6   7   8   9

MSI/LSI FUNCTIONS
FUNCTIONAL INDEX/SELECTION GUIDE

PROGRAMMABLE READ-ONLY MEMORIES (PROM'S)[†]

DESCRIPTION	ORGANI-ZATION	TYPE OF OUTPUT	TYPICAL ADDRESS TIME	TYPICAL ENABLE TIME	TYP POWER DISSIPATION PER BIT	DEVICE TYPE AND PACKAGE			
						−55°C to 125°C		0°C to 70°C	
4096-BIT ARRAYS	512 X 8	3-State	55 ns	20 ns	0.14 mW	SN54S472	J	SN74S472	J, N
	512 X 8	O-C	55 ns	20 ns	0.14 mW	SN54S473	J	SN74S473	J, N
	512 X 8	3-State	55 ns	20 ns	0.14 mW	SN54S474	J, W	SN74S474	J, N
	512 X 8	O-C	55 ns	20 ns	0.14 mW	SN54S475	J, W	SN74S475	J, N
2048-BIT ARRAYS	256 X 8	O-C	50 ns	20 ns	0.24 mW	SN54S470	J	SN74S470	J, N
	256 X 8	3-State	50 ns	20 ns	0.27 mW	SN54S471	J	SN74S471	J, N
1024-BIT ARRAYS	256 X 4	3-State	40 ns	15 ns	0.49 mW	SN54S287	J, W	SN74S287	J, N
	256 X 4	O-C	40 ns	15 ns	0.49 mW	SN54S387	J, W	SN74S387	J, N
512-BIT ARRAYS	64 X 8	O-C	50 ns	47 ns	0.6 mW	SN54186	J, N	SN74186	J, N
256-BIT ARRAYS	32 X 8	O-C	29 ns	28 ns	1.3 mW	SN54188A	J, W	SN74188A	J, N
	32 X 8	O-C	25 ns	12 ns	1.56 mW	SN54S188	J, W	SN74S188	J, N
	32 X 8	3-State	25 ns	12 ns	1.56 mW	SN54S288	J, W	SN74S288	J, N

MICROPROCESSOR CONTROLLERS AND SUPPORT FUNCTIONS

DESCRIPTION	SYSTEM APPLICATION	TYP TOTAL POWER DISSIPATION	DEVICE TYPE AND PACKAGE				PAGE NO.
			−55°C to 125°C		0°C to 70°C		
SYSTEM CONTROLLERS	8080A	700 mW			SN74S428 (TIM8228)	N	7-514
	8080A	700 mW			SN74S438 (TIM8238)	N	7-514
	Universal	450 mW	SN54S482	J	SN74S482	J, N	†
REGISTERS	TMS 9900	110 mW	SN54LS259	J, W	SN74LS259 (TIM9906)	J, N	7-376
	MOS	210 mW	SN54LS363*	J	SN74LS363*	J, N	7-467
		210 mW	SN54LS364*	J	SN74LS364*	J, N	7-467
MULTI-MODE LATCHES	8080A	410 mW	SN54S412	J, W	SN74S412 (TIM8212)	J, N	7-502
TRANSCEIVERS AND BUS DRIVERS		625 mW	SN54S226*	J, W	SN74S226*	J, N	7-345
		207 mW	SN54LS245*	J	SN74LS245*	J, N	7-349
TRANSCEIVERS AND BUS DRIVERS (SSI)		98 mW	SN54LS240	J	SN74LS240	J, N	6-83
		450 mW	SN54S240	J	SN74S240	J, N	6-83
		100 mW	SN54LS241	J	SN74LS241	J, N	6-83
		538 mW	SN54S241	J	SN74S241	J, N	6-83
		128 mW	SN54LS242	J, W	SN74LS242	J, N	6-87
		128 mW	SN54LS243	J, W	SN74LS243	J, N	6-87
		100 mW	SN54LS244	J	SN74LS244	J, N	6-83
CLOCK ELEMENTS	TMS 9900	669 mW			SN74LS362 (TIM9904)*	J, N	7-460
	8080A	719 mW			SN74LS424 (TIM8224)	J, N	7-507
LOGIC ELEMENTS	TMS 9900	190 mW	SN54148	J, W	SN74148 (TIM9907)	J, N	7-151
	TMS 9900	35 mW	SN54LS251	J, W	SN74LS251 (TIM9905)	J, N	7-362
	TMS 9900	63 mW	SN54LS348*	J, W	SN74LS348 (TIM9908)*	J, N	7-448

*New product in development as of October 1976.
[†]See Bipolar Microcomputer Components Data Book, LCC4270.

TEXAS INSTRUMENTS

TTL INTEGRATED CIRCUITS MECHANICAL DATA

ORDERING INSTRUCTIONS

Electrical characteristics presented in this data book, unless otherwise noted, apply for circuit type(s) listed in the page heading regardless of package. The availability of a circuit function in a particular package is denoted by an alphabetical reference above the pin-connection diagram(s). These alphabetical references refer to mechanical outline drawings shown in this section.

Factory orders for circuits described in this catalog should include a four-part type number as explained in the following example.

EXAMPLE: SN 54LS75 J −00

1. Prefix

MUST CONTAIN TWO OR THREE LETTERS
(From Individual Data Sheet)

RSN	Radiation-Hardened Circuit
SN	Standard Prefix
SNM	Mach IV, Level I
SNC	Mach IV, Level III
SNH	Mach IV, Level IV
SNJ	JAN Processed

2. Unique Circuit Description

MUST CONTAIN FOUR TO EIGHT CHARACTERS
(From Individual Data Sheet)

Examples:
5410
74H10
54S112
54L78
74LS295A
74188A

3. Package

MUST CONTAIN ONE OR TWO LETTERS
J, JD, N, T, W
(From Pin-Connection Diagram on Individual Data Sheet)

4. Instructions (Dash No.)

MUST CONTAIN TWO NUMBERS
(From Dash No. Column of Following Table)

PACKAGES	FORMED LEADS	SOLDER-DIPPED LEADS	INSULATOR	CARRIER	ORDER DASH NO.
METAL FLAT PACKAGES					
T	No	No	No	†	00
T	Yes	No	Yes	†	01
T	No	No	No	Mech-Pak	02
T	No	No	Yes	Mech-Pak	03
T	Yes	No	No	Mech-Pak	04
T	Yes	No	Yes	Mech-Pak	05
T	No	No	Yes	†	06
T	Yes	No	No	†	07
T	No	Yes	No	†	10
T	Yes	Yes	Yes	†	11
T	No	Yes	No	Mech-Pak	12
T	No	Yes	Yes	Mech-Pak	13
T	Yes	Yes	No	Mech-Pak	14
T	Yes	Yes	Yes	Mech-Pak	15
T	No	Yes	Yes	†	16
T	Yes	Yes	No	†	17
CERAMIC FLAT PACKAGES					
W	No	No	N/A	†	00
W	No	Yes	N/A	†	10
DUAL-IN-LINE PACKAGES					
J, JD, N	No	No	N/A	†	00
N	No	Yes	N/A	†	10

†These circuits are shipped in one of the carriers shown below. Unless a specific method of shipment is specified by the customer (with possible additional posts), circuits will be shipped in the most practical carrier. Please contact your TI sales representative for the method that will best suit your particular needs.

Flat (T, W)	Dual-in-line ((J, JD, N)
—Barnes Carrier	—Slide Magazines
—Milton Ross Carrier	—A-Channel Plastic Tubing
	—Barnes Carrier (N only)
	—Sectioned Cardboard Box
	—Individual Plastic Box

TEXAS INSTRUMENTS

TTL INTEGRATED CIRCUITS MECHANICAL DATA

N plastic dual-in-line packages

These dual-in-line packages consist of a circuit mounted on a 14-, 16-, 20-, or 28-lead frame and encapsulated within an electrically nonconductive plastic compound. The compound will withstand soldering temperature with no deformation and circuit performance characteristics remain stable when operated in high-humidity conditions. The packages are intended for insertion in mounting hole rows on 0.300 (7,62) or 0.600 (15,24) centers. Once the leads are compressed and inserted, sufficient tension is provided to secure the package in the board during soldering. Leads require no additional cleaning or processing when used in soldered assembly.

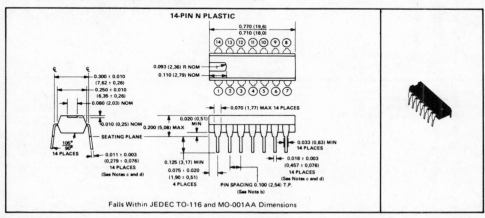

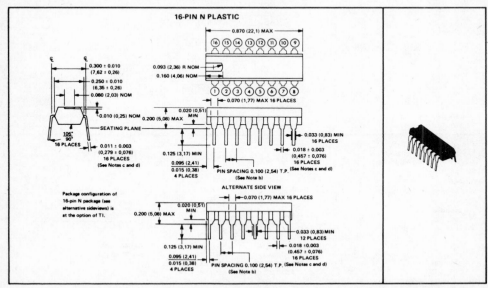

TEXAS INSTRUMENTS

54/74 FAMILIES OF COMPATIBLE TTL CIRCUITS

PIN ASSIGNMENTS (TOP VIEWS)

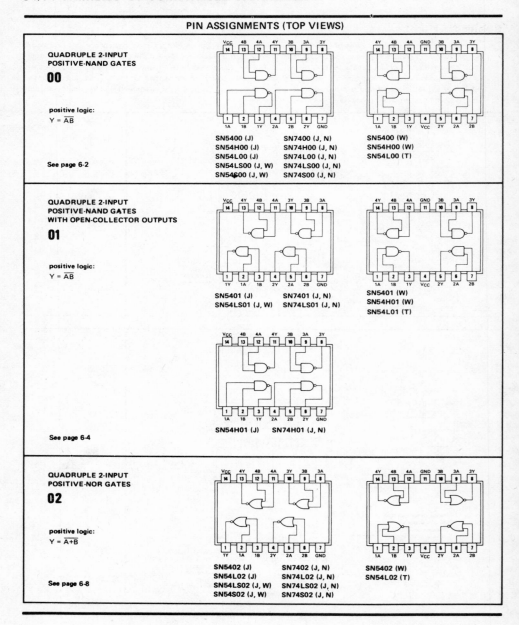

QUADRUPLE 2-INPUT POSITIVE-NAND GATES

00

positive logic:
$Y = \overline{AB}$

See page 6-2

SN5400 (J)   SN7400 (J, N)   SN5400 (W)
SN54H00 (J)   SN74H00 (J, N)   SN54H00 (W)
SN54L00 (J)   SN74L00 (J, N)   SN54L00 (T)
SN54LS00 (J, W)   SN74LS00 (J, N)
SN54S00 (J, W)   SN74S00 (J, N)

QUADRUPLE 2-INPUT POSITIVE-NAND GATES WITH OPEN-COLLECTOR OUTPUTS

01

positive logic:
$Y = \overline{AB}$

SN5401 (J)   SN7401 (J, N)   SN5401 (W)
SN54LS01 (J, W)   SN74LS01 (J, N)   SN54H01 (W)
                                       SN54L01 (T)

SN54H01 (J)   SN74H01 (J, N)

See page 6-4

QUADRUPLE 2-INPUT POSITIVE-NOR GATES

02

positive logic:
$Y = \overline{A+B}$

See page 6-8

SN5402 (J)   SN7402 (J, N)   SN5402 (W)
SN54L02 (J)   SN74L02 (J, N)   SN54L02 (T)
SN54LS02 (J, W)   SN74LS02 (J, N)
SN54S02 (J, W)   SN74S02 (J, N)

TEXAS INSTRUMENTS

54/74 FAMILIES OF COMPATIBLE TTL CIRCUITS

PIN ASSIGNMENTS (TOP VIEWS)

AND-GATED J-K MASTER-SLAVE FLIP-FLOPS WITH PRESET AND CLEAR

72

FUNCTION TABLE

INPUTS					OUTPUTS	
PRESET	CLEAR	CLOCK	J	K	Q	Q̄
L	H	X	X	X	H	L
H	L	X	X	X	L	H
L	L	X	X	X	H*	H*
H	H	⎍	L	L	Q₀	Q̄₀
H	H	⎍	H	L	H	L
H	H	⎍	L	H	L	H
H	H	⎍	H	H	TOGGLE	

positive logic: $J = J1 \cdot J2 \cdot J3; \ K1 \cdot K2 \cdot K3$

See pages 6-46, 6-50, and 6-54

SN5472 (J)   SN7472 (J, N)   SN5472 (W)
SN54H72 (J)   SN74H72 (J, N)   SN54H72 (W)
SN54L72 (J)   SN74L72 (J, N)   SN54L72 (T)

NC—No internal connection

DUAL J-K FLIP-FLOPS WITH CLEAR

73

'73, 'H73, 'L73
FUNCTION TABLE

INPUTS				OUTPUTS	
CLEAR	CLOCK	J	K	Q	Q̄
L	X	X	X	L	H
H	⎍	L	L	Q₀	Q̄₀
H	⎍	H	L	H	L
H	⎍	L	H	L	H
H	⎍	H	H	TOGGLE	

'LS73
FUNCTION TABLE

INPUTS				OUTPUTS	
CLEAR	CLOCK	J	K	Q	Q̄
L	X	X	X	L	H
H	↓	L	L	Q₀	Q̄₀
H	↓	H	L	H	L
H	↓	L	H	L	H
H	↓	H	H	TOGGLE	
H	H	X	X	Q₀	Q̄₀

See pages 6-46, 6-50, 6-54, and 6-56

SN5473 (J, W)   SN7473 (J, N)
SN54H73 (J, W)   SN74H73 (J, N)
SN54L73 (J, T)   SN74L73 (J, N)
SN54LS73 (J, W)   SN74LS73 (J, N)

DUAL D-TYPE POSITIVE-EDGE-TRIGGERED FLIP-FLOPS WITH PRESET AND CLEAR

74

FUNCTION TABLE

INPUTS				OUTPUTS	
PRESET	CLEAR	CLOCK	D	Q	Q̄
L	H	X	X	H	L
H	L	X	X	L	H
L	L	X	X	H*	H*
H	H	↑	H	H	L
H	H	↑	L	L	H
H	H	L	X	Q₀	Q̄₀

See pages 6-46, 6-50, 6-54, and 6-56

SN5474 (J)   SN7474 (J, N)   SN5474 (W)
SN54H74 (J)   SN74H74 (J, N)   SN54H74 (W)
SN54L74 (J)   SN74L74 (J, N)   SN54L74 (T)
SN54LS74A (J, W)   SN74LS74A (J, N)
SN54S74 (J, W)   SN74S74 (J, N)

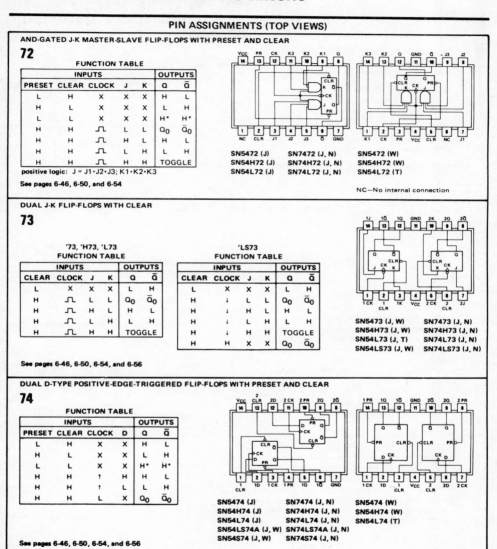

See explanation of function tables on page 3-8.
*This configuration is nonstable; that is, it will not persist when preset and clear inputs return to their inactive (high) level.

TTL MSI

TYPES SN54LS138, SN54LS139, SN54S138, SN54S139, SN74LS138, SN74LS139, SN74S138, SN74S139 DECODERS/DEMULTIPLEXERS

BULLETIN NO. DL-S 7611804, DECEMBER 1972–REVISED OCTOBER 1976

- Designed Specifically for High-Speed:
  Memory Decoders
  Data Transmission Systems

- 'S138 and 'LS138 3-to-8-Line Decoders Incorporate 3 Enable Inputs to Simplify Cascading and/or Data Reception

- 'S139 and 'LS139 Contain Two Fully Independent 2-to-4-Line Decoders/ Demultiplexers

- Schottky Clamped for High Performance

TYPE	TYPICAL PROPAGATION DELAY (3 LEVELS OF LOGIC)	TYPICAL POWER DISSIPATION
'LS138	22 ns	32 mW
'S138	8 ns	245 mW
'LS139	22 ns	34 mW
'S139	7.5 ns	300 mW

description

These Schottky-clamped TTL MSI circuits are designed to be used in high-performance memory-decoding or data-routing applications requiring very short propagation delay times. In high-performance memory systems these decoders can be used to minimize the effects of system decoding. When employed with high-speed memories utilizing a fast-enable circuit the delay times of these decoders and the enable time of the memory are usually less than the typical access time of the memory. This means that the effective system delay introduced by the Schottky-clamped system decoder is negligible.

The 'LS138 and 'S138 decode one-of-eight lines dependent on the conditions at the three binary select inputs and the three enable inputs. Two active-low and one active-high enable inputs reduce the need for external gates or inverters when expanding. A 24-line decoder can be implemented without external inverters and a 32-line decoder requires only one inverter. An enable input can be used as a data input for demultiplexing applications.

SN54LS138, SN54S138 . . . J OR W PACKAGE
SN74LS138, SN74S138 . . . J OR N PACKAGE
(TOP VIEW)

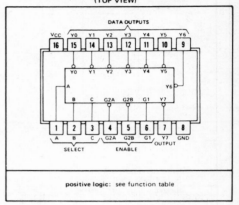

positive logic: see function table

SN54LS139, SN54S139 . . . J OR W PACKAGE
SN74LS139, SN74S139 . . . J OR N PACKAGE
(TOP VIEW)

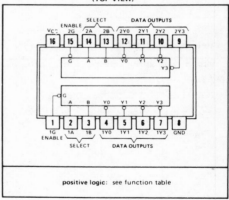

positive logic: see function table

The 'LS139 and 'S139 comprise two individual two-line-to-four-line decoders in a single package. The active-low enable input can be used as a data line in demultiplexing applications.

All of these decoders/demultiplexers feature fully buffered inputs each of which represents only one normalized Series 54LS/74LS load ('LS138, 'LS139) or one normalized Series 54S/74S load ('S138, 'S139) to its driving circuit. All inputs are clamped with high-performance Schottky diodes to suppress line-ringing and simplify system design. Series 54LS and 54S devices are characterized for operation over the full military temperature range of −55°C to 125°C; Series 74LS and 74S devices are characterized for 0°C to 70°C industrial systems.

TEXAS INSTRUMENTS

TTL
MSI

TYPES SN54181, SN54LS181, SN54S181,
SN74181, SN74LS181, SN74S181
ARITHMETIC LOGIC UNITS/FUNCTION GENERATORS
BULLETIN NO. DL-S 7611831, DECEMBER 1972 — REVISED OCTOBER 1976

- Full Look-Ahead for High-Speed Operations on Long Words

- Input Clamping Diodes Minimize Transmission-Line Effects

- Darlington Outputs Reduce Turn-Off Time

- Arithmetic Operating Modes:
  Addition
  Subtraction
  Shift Operand A One Position
  Magnitude Comparison
  Plus Twelve Other Arithmetic Operations

- Logic Function Modes:
  Exclusive-OR
  Comparator
  AND, NAND, OR, NOR
  Plus Ten Other Logic Operations

SN54181, SN54LS181, SN54S181 . . . J OR W PACKAGE
SN74181, SN74LS181, SN74S181 . . . J OR N PACKAGE
(TOP VIEW)

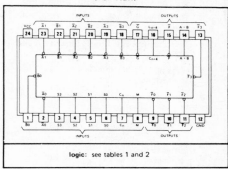

logic: see tables 1 and 2

TYPICAL ADDITION TIMES

NUMBER OF BITS	ADDITION TIMES			PACKAGE COUNT		CARRY METHOD BETWEEN ALU's
	USING '181 AND '182	USING 'LS181 AND '182	USING 'S181 AND 'S182	ARITHMETIC/ LOGIC UNITS	LOOK-AHEAD CARRY GENERATORS	
1 to 4	24 ns	24 ns	11 ns	1		NONE
5 to 8	36 ns	40 ns	18 ns	2		RIPPLE
9 to 16	36 ns	44 ns	19 ns	3 or 4	1	FULL LOOK-AHEAD
17 to 64	60 ns	68 ns	28 ns	5 to 16	2 to 5	FULL LOOK-AHEAD

description

The '181, 'LS181, and 'S181 are arithmetic logic units (ALU)/function generators that have a complexity of 75 equivalent gates on a monolithic chip. These circuits perform 16 binary arithmetic operations on two 4-bit words as shown in Tables 1 and 2. These operations are selected by the four function-select lines (S0, S1, S2, S3) and include addition, subtraction, decrement, and straight transfer. When performing arithmetic manipulations, the internal carries must be enabled by applying a low-level voltage to the mode control input (M). A full carry look-ahead scheme is made available in these devices for fast, simultaneous carry generation by means of two cascade-outputs (pins 15 and 17) for the four bits in the package. When used in conjunction with the SN54182, SN54S182, SN74182, or SN74S182, full carry look-ahead circuits, high-speed arithmetic operations can be performed. The typical addition times shown above illustrate the little additional time required for addition of longer words when full carry look-ahead is employed. The method of cascading '182 or 'S182 circuits with these ALU's to provide multi-level full carry look-ahead is illustrated under typical applications data for the '182 and 'S182.

If high speed is not of importance, a ripple-carry input (C_n) and a ripple-carry output (C_{n+4}) are available. However, the ripple-carry delay has also been minimized so that arithmetic manipulations for small word lengths can be performed without external circuitry.

TEXAS INSTRUMENTS

TTL
MSI

TYPES SN54LS245, SN74LS245
OCTAL BUS TRANSCEIVERS WITH 3-STATE OUTPUTS

BULLETIN NO. DL-S 7612471, OCTOBER 1976

- Bi-directional Bus Transceiver in a High-Density 20-Pin Package
- 3-State Outputs Drive Bus Lines Directly
- P-N-P Inputs Reduce D-C Loading on Bus Lines
- Hysteresis at Bus Inputs Improve Noise Margins
- Typical Propagation Delay Times, Port-to-Port . . . 12 ns
- Typical Enable/Disable Times . . . 17 ns

SN54LS245 . . . J PACKAGE
SN74LS245 . . . J OR N PACKAGE
(TOP VIEW)

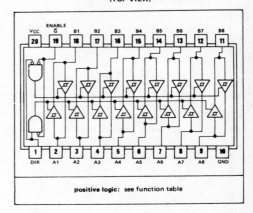

positive logic: see function table

TYPE	I_{OL} (SINK CURRENT)	I_{OH} (SOURCE CURRENT)
SN54LS245	12 mA	−12 mA
SN74LS245	24 mA	−15 mA

description

These octal bus transceivers are designed for asynchronous two-way communication between data buses. The control function implementation minimizes external timing requirements.

The device allows data transmission from the A bus to the B bus or from the B bus to the A bus depending upon the logic level at the direction control (DIR) input. The enable input ($\overline{G}$) can be used to disable the device so that the buses are effectively isolated.

The SN54LS245 is characterized for operation over the full military temperature range of −55°C to 125°C. The SN74LS245 is characterized for operation from 0°C to 70°C.

schematics of inputs and outputs

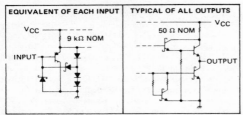

FUNCTION TABLE

ENABLE $\overline{G}$	DIRECTION CONTROL DIR	OPERATION
L	L	B data to A bus
L	H	A data to B bus
H	X	Isolation

H = high level, L = low level, X = irrelevant

absolute maximum ratings over operating free-air temperature range (unless otherwise noted)

Supply voltage, V_{CC} (see Note 1) . . . . 7 V
Input voltage . . . . 7 V
Operating free-air temperature range: SN54LS245 . . . . −55°C to 125°C
                                        SN74LS245 . . . . 0°C to 70°C
Storage temperature range . . . . −65°C to 150°C

NOTE 1: Voltage values are with respect to network ground terminal.

TEXAS INSTRUMENTS

Index

A

abstract stream in computer logic, 23
accumulator, 99, 100, 158
ACIA (asynchronous communications inter-
    face adapter), 48, 151
active–HI input, 50
active–LO input, 50, 96, 100
active pull-up, 39
adder, binary, 70–75
adder, parallel, 160–164
adder unit, 36, 52, 65
ADD instruction in ALU, 159–160
ADD instruction, micro-instruction sequence
    for, 19–22, 166–169
addition using 2s complement, 156, 157
address bus, 19, 142, 152
address decoding logic, 50, 142, 146
address field of micro-instruction, 173, 174
addressing capability, 19, 49, 182
addressing computer memory, 48–50
ad hoc design, 140
adjacent square K-map, 74–76
algebraic manipulation, 33, 72, 75
algebra of propositions, 30
ALGOL, 15
ALU, 14, 19, 65, 70
ALU structure, 156–164
ALU structure, detailed, 159–160
amplification, 83
amplification property of transistor, 38
analogue computer, 11
analogue-to-digital (A/D) converter, 44
Analytical Engine, 11
AND gate, see gates
AND operation in ALU, 160
applications software, 9, 10, 179
arc, 64
arithmetic and logic unit, see ALU
arithmetic operations, 156–164
ASCII code, 115

assembly language, 10
assignment of binary codes to internal
    states, 109, 126
associative laws (of Boolean algebra), 32
associative memory, 183–184
asynchronous input, 96, 99, 106
asynchronous operations, 15
asynochrnous sequential logic circuit, 57,
    101, 107, 134, 147
ATLAS computer, 14, 182
automata, 25
avalanche effect, 102

B

Babbage, Charles, 11, 12
Backus, John, 15
Bardeen, J., 14, 36, 187
base of transistor, 37, 38, 42
basic rules of Boolean algebra, 30, 31
BCD to Gray code conversion, 78–82
BCD to seven-segment LED decoder, 119–
    125
binary adder, 70–75, 94, 158–159
binary-coded-decimal, see BCD
binary number system, 26
bipolar transistor (see also transistor), 36,
    37
bistable (see also flip-flop), 26, 91, 92
bit level in computer logic, 24, 144
bit, 9
Boole, George, 30, 188
Boolean algebra, 22, 23, 72, 83, 86
    duality in, 33
    postulates of, 31
    theorems of, 30–34
    violation of laws of, 132
Boolean expression, 54, 76, 78, 86
Boolean functions of two variables, 34, 35
Boolean operates, 58–60

Boolean variable, 30
bootstrap, 51, 150
borrow, 160
Brattain, W. H., 14, 36, 187
buffer, 138, 152
Burks, A. W., 13, 187
bus
    CPU, 19, 152, 164, 165
    address, 19, 142, 152
    data, 19, 142, 152
byte, 149

C

cache memory, 184
canonical expansion, 112
canonical form, 76, 77
canonical product term, 86, 147
carry
    carry-in,
        carry-out, 70–75
carry look-ahead, 160–164
carry look-ahead generator, 65, 162–163
central processing unit, *see* CPU
ceramic IC package, 45, 48
characteristic equation, 102, 104, 105
chip, 36, 185, 186
chip enable, 146
chip select, 49
circuit board layout, 83
circuit symbols for flip-flops, 95
circuit symbols for gates, 29
classification of computers, 151
classification of logic circuits, 57
clear (CLR) input, 96, 98, 99
clock, 15, 22, 92–102, 139, 140
clock frequency, 22, 94
clock input (CK), 94, 96
clock pulse, 98, 99, 100
clocked flip-flop, 94
closure rule (*see also* Boolean algebra), 31
CMOS-TTL interface, 137, 138
COBOL, 15
code converter, 65
collector of transistor, 37, 38, 42
combinational logic, 22, 23, 25, 26, 57, 65
    design, 56–90
    design, steps in, 68
common-emitter configuration, 38
commutative operations (*see also* Boolean algebra), 31, 32
comparator, 65, 160
comparison of logic families, 43
compiler, 14, 15, 182
complementary MOS (CMOS), 42, 43
complementation of sets, 60
completeness (*see also* Boolean algebra), 30

computers
    analogue, 11
    digital, 11
    generations of, 13–17
    history of, 11–17
    maintenance, 16, 17, 151
    microcomputer, 17, 50, 149
    microprocessor, 16, 17, 19, 36, 46, 48, 52, 137, 140–142, 149, 151
    minicomputer, 16, 17, 151
computer architecture, 18, 22, 178–186
computer design, 150–177
computer hardware, 10, 11, 14, 178, 185, 186
computer logic, definition, 11, 18
computer software, 9–11, 15, 178, 185, 186
condition codes register (CC), *see* registers
conditional branching, 175, 176
content-addressable memory, 184
control field of micro-instruction, 173, 174
control memory, 173–177
    address register (CMAR), 173–176
    data register (CMDR), 173–177
control signals, 164–177
control unit, 19–21
control unit, microprogrammed version, 175, 176
control unit, structure of, 164–177
counter, 100, 102–109
counters, types of
    asynchronous, 107–109
    down, 103
    modulo–8, 102–108
    modulo–N, 115–119, 170
    re-cycling, 103
    ring, 103, 170
    self-stopping, 106
    synchronous, 103–108
CPU, 14, 15, 18, 19, 50, 52, 66
    bus, 19, 152, 164, 165
    structure, 151–154
    structure, detailed, 164
current-sinking, 135
current-sourcing, 135
cyclic counter, 116

D

data bus, 19, 142, 152
data sheet (*see also* Appendix, 193–206), 45–48, 78, 94, 134, 137, 140
data transfer, parallel, serial, 154, 155
DCTL, 14
decoder, 65, 155, 156, 170, 171
decoding of states, 100, 128
decoupling capacitor, 94
delay, 25

de Morgan's laws (*see also* Boolean algebra), 31–33, 56, 72
de-multiplexer, 156
Denning, P. J., 184, 188, 191, 192
depletion-mode FET, 41
design complexity, 100
designator of IC Package, 47
D flip-flop (*see also* flip-flop) 38, 95–98
digital clock, 102, 116, 119
digital computer, 11
Digital Equipment Corporation (DEC), 16
digital IC, 44, 48–52
digital-to-analogue (D/A) converter, 44
DIL package, 45–48
diode logic, 83
diode transistor logic (DTL), 14, 36
direct-coupled transistor logic (DCTL), 14
discrete transistor technology, 14
distinctive shapes (for logic symbols), 28, 53
distributed computer system, 48
distributive operations (*see also* Boolean algebra), 31, 32
divide-by-two counter, 108
divide (DIV) instruction, 151, 156
don't care conditions, 78–80, 98, 111
doping, 37
drain (of MOSFET), 42
drawing templates (for logic symbols), 28
drive capability, 137, 138
DTL, 14, 36
D-type input, 96–98
dual forms in Boolean algebra, 33
dual IC package, 47
duality in Boolean algebra, 33
duration of pulse, 93
duty cycle, 94
dynamic hazard, 174
dynamic microprogramming, 173
dynamic RAM, 51

E

Eckert, J. P., 12
edge-triggered flip-flop, 97, 98
edge-triggered operation, 93, 94, 133
editor, 9
EDSAC, 14
EDVAC, 12, 14
Electric Tabulating System, 11
electric switching elements, 36
electrons, 37, 38, 41, 42
elimination of variables, 74
emitter-coupled logic (ECL), 41–43
emitter of transistor, 37, 38, 42
emulation, 178
enable input, 97, 156

energy required for switching operation, 43
enhancement-mode FET, 41, 42
ENIAC, 11, 12
EOR, 54, 55, 67, 72, 73, 81, 82, 114
    gate (*see also* gates), 34, 82
    operation in ALU, 160
erasable PROM (EPROM), 51
essential hazard, 134
essential PI, 77, 78, 86, 89
even parity, 113
excitation matrix, 101
exclusive–OR (*see* EOR)
exponent, 156
external pull-up resistor, 137, 138

F

fabrication, 36, 42, 43
factorisation, 68, 84, 131, 134
Fairchild, 16
falling edge, 93
fan-in, 83
fanout (*see also* gates), 39, 42, 83, 84, 134–138
feedback, 65, 83, 90, 100, 147, 148
fetch machine-code instruction, 20–22
field-effect transistor (FET), 41, 42
field-programmable ROM, 51
finite-state machine (*see* FSM)
first-generation digital computer, 13, 17
flat IC package, 48
flip-flop, 26, 52, 58, 90–100
    characteristic equations of, 102
    S–R, 28, 29, 90–96
    D, 28, 95–98
    J–K, 28, 95–102
floating point operation, 156
flow table, 65, 101
FORTRAN, 15
fourth-generation digital computer, 17
FSM, 25, 63, 124
full-adder, 71–75
function table, 98, 99

G

gain of transistor, 39
gate (of MOSFET), 42
gate count, 85, 86
gate loading, 135–137
gated latch, 97
gates, 26–30, 52–56
    AND, 68, 83, 131
    EOR, 34, 82
    NAND, 40, 68, 83–86, 131
    NOR, 68

gates (*continued*)
    NOT, 39, 68, 83, 131
    OR, 68, 83, 131
    physical realisation of, 36–43
    propagation delay of, 68, 94
generate term, 162
glitch, 93
GND, 45, 47, 96, 135, 136
Goldstine, H. H., 13, 187
Gray code, 62, 63, 78–82, 103

H

half-adder, 71–73
hard-wired logic, 149
hard-wired method of computer design, 173
hazard, 109, 131–134, 147
hazard elimination, 134
hex IC package, 47
hexadecimal, 142
HI voltage level, 38, 39, 52, 53, 135–137
HI voltage range for TTL, 137
high-impedance, 44, 165
high-level programming, 14, 178–182
high-speed (H) TTL logic, 41, 44
history of computers, 11–17
hit rate, 184
holes, 37, 41, 42
Hollerith, H., 11
horizontal micro-instruction, 176–177
host, 170
Huffman, D. A., 65
Huntington, E. V., 30, 189
Huntington's postulates, 30

I

IAS computer, 13
IBM, 12, 16
IC (integrated circuit) (*see also* Appendix
    193–206), 16, 23, 36, 44–56, 68–70
    board layout, 69
    package designator, 47, 48
    package count, 83–86
    packaging, 45–48
    packages
        7400 (quad 2-input NAND), 47, 70,
            85, 86
        7404 (hex inverter), 85, 86, 141, 143
        7401 (triple 3-input NAND), 70, 85,
            86, 141, 143
        7414 (hex Schmitt inverter), 138
        7420 (dual 4-input NAND), 85, 86,
            141, 143
        7448 (BCD to seven-segment LED
            decoder), 124
        7473 (dual J-K), 96, 99

IC (integrated circuit) (*continued*)
    packages (*continued*)
        7474 (dual D), 98, 141, 143
        74141 (monostable), 141, 143
        74123 (dual monostable), 139
        74148 (8-line to 3-line priority
            encoder), 112
        74160 (synchronous mod-10 coun-
            ter), 119
        74180 (parity generator/checker),115
        74181 (ALU), 160–163
        74182 (carry look-ahead generator),
            162, 163
        74244 (octal bus driver), 153
        74245 (octal transceiver), 152, 153
        8T95 (hex tri-state buffer), 141, 143,
            152
    socket, 69
identity elements (in Boolean algebra), 31
implementation of logic circuits, 35, 36,
    45–48
implicit strengths of Boolean operators, 32
increment operation in ALU, 160
index register, 15
indirect addressing, 15, 20
infix form, 180
input/output (I/O), 19, 113, 181
instruction decoder, 171
instruction register (IR) (*see* registers)
instruction register, use of in control units,
    171–176
integrated circuit (*see* IC)
integrated injection logic (I²L), 41–43
INTEL, 4004, 8080 microprocessors, 16
interconnections, 68–70, 83–85
interface, 11
interface circuit, 140–143
interface ICs, 44, 48
interfacing of logic families, 136–138
internal state, 24–26, 63–65, 91, 92, 100–
    104, 124, 126
internal state reduction, 109
interrupt, 110
interrupt handler, 152
interrupt mechanism, 181
intersection of sets, 60
interval between pulses, 94
intuitive design methods, 109
inverse of elements (in Boolean algebra),
    31
inversion of logic signal, 53–54
inverting action of transistor, 39

J

J-K flip-flop (*see also* flip-flop), 28, 95–
    102

K

Karnaugh, M., 189
Karnaugh map (*see* K-map)
Kilburn, T., 14
K-map, 60–63, 73–78, 83, 86, 89, 101, 102, 105

L

large-scale integration (*see* LSI)
latch, 97, 98
layers in computer architecture, 10, 11, 22, 23, 178, 179
leading edge, 93, 98, 139
leakage current, 38
level of integration, 16
level-sensitive flip-flop, 93, 94
levels in computer logic, 23, 24, 144
levels of logic (delays), 68, 83–85
linear IC, 44
line driving/receiving, 48
LO voltage level, 38–39, 52, 53, 135–137
LO voltage range for TTL, 137
logic
    building blocks, 22–24
    circuit analysis, 140–143
    circuits, 11, 24
    diagrams, 54, 55
    families, 14, 16, 34–45, 136–138
    level transition, 93
    negative, 52–56
    positive, 52–56
    symbols, 29, 52–56, 95
logical expressions, 30
logical operations in ALU, 158–160
low power consumption, 138
low-power Schottky (LS), 41, 44, 136, 137
low-power TTL (L), 41, 44, 137
LSI, 16, 23, 36, 41–46, 52, 58, 185

M

machine-code instruction, 10, 19–22, 153, 154, 170
machine cycle, 22, 169–171
magnetic disc, tape, 151
mainframe computer, 15, 16, 17
man-machine interface, 11
mantissa, 156
mask-programmable ROM, 51
master-slave J–K flip-flop, 98–100
Mauchly, J. W., 11, 12
McCluskey, E., 86, 189
Mealy, G. H., 65, 190
medium-scale integration (*see* MSI)
memory, 13, 14, 18, 19, 65, 91, 100
    types of, 48–51

memory address register (MAR) (*see* registers)
memory data registers (MDR) (*see* registers)
memory-mapped I/O, 19, 48
memory space in VM, 182
metal-oxide- semiconductor (MOS) technique, 42
microcomputer, 17, 50, 149
microcomputer, single-chip, 149
micro-instruction (micro-operation)
    encoder, 171
    flow-chart, 172
    format, 175
    horizontal, 176
    vertical, 177
microprocessor, 16–17, 19, 36, 46, 48, 52, 137, 140–142, 149
microprogrammed control unit, 175, 176
microprogramming, 173–177
minicomputer, 16, 17, 151
minimisation, 60, 147, 148
    criteria, 24, 25, 144
    notes on, 83–86
minterm, 77, 86
mode control, 159–160
mode field of micro-instruction, 175, 176
modulo-8 counter, 102–108
modulo-N counter, 115–119, 170
monitor program, 51
monolithic structure, 36
monostable, 139, 140, 143
Moore, G. F., 65, 189
Moore/Mealy model, 65, 90, 100, 147
Moore school, 11
MOS, 137
    CMOS, 137–138
    NMOS, 137
    PMOS, 137
MOSFET, 42
MOS–TTL interface, 137
Motorola
    6800 microprocessor, 16, 142, 151
    6820 PIA, 48, 151
    6850 ACIA, 48, 151
MSI, 10, 23, 28, 36, 41–47, 52, 58, 66, 67
multi-level machine, 179
multi-output function, 145
multiple-emitter transistor, 40
multiple-output problem, 68, 78
multiplexer (MUX), 66, 144, 145
multiplicand, 158
multiplier, 158
multiplier unit, 36
multiply instruction, 151, 156, 158
multiprogramming, 181, 182
multi-variable hazard, 134

N

name space in VM, 182
NAND gate implementation (*see also* gates), 40
Naur, Peter, 15
n-channel FET, 41, 42
negative logic, 52–56
negative pulse, 93
NMOS transistor, 42, 43
node, 64, 100
noise, 93, 94
    immunity, 39
    margin, 137, 138
nominal gate functions, 53
non-saturated mode transistors, 41
non-volatile memory, 50
NOR gate (*see also* gates), 68, 83, 181
normalised fanout, 136
NOT gate implementation (*see also* gates), 39
n-p-n transistor, 37, 41
n-type doping, impurities, 37

O

octal IC packages, 117
odd parity, 113
OFF state of transistor, 38, 39
one (1)-level state, 14, 182
one-shot, 139
ON state of transistor, 38, 39, 93
open-collector output, 44
operating system, 9, 10, 179
operational amplifier, 44
optimised instruction set, 185
OR gate (*see also* gates), 68, 83, 131
OR operation in ALU, 160
ordering ICs, 47, 48
organisation of memory chips, 49
output characteristic of transistor, 38, 39
output configurations of TTL logic, 44
output stage of TTL gate, 40

P

packing density, 41–43, 48, 151
page-fault interrupt, 184
paging, 182–184
paper tape reader, 139–143
parallel addition, 70, 71, 160–164
parallelism, 176, 177, 184, 185
parity generator/checker, 66, 113–115
partial address decoding, 142
pattern recogniser, 124, 126–128
p-channel FET, 41, 42
PDP-11 minicomputer, 160
perfect induction, method of, 33

peripheral device, 18, 48, 110
phase-locked loop, 44
physical stream in computer logic, 23
PI (prime implicant), 77, 78, 86, 89, 90
PIA (peripheral interface adapter), 48, 151
pin configuration, 49, 50
pin count, 46, 84
pins in IC package, 45–47
pipelining, 184, 185
plastic IC package, 45, 48
PLA (programmable logic array), 147, 148, 149
PMOS transistor, 42, 43
p-n junction, 37
p-n-p transistor, 37, 41
point-to-point wiring diagram, 69
positive logic, 52–56, 90, 93
positive pulse, 93
positive-edge-triggered flip-flop, 98, 119
positive–NOR, 52
postulates of Boolean algebra, 30, 59, 60
power consumption of logic families, 41–44
preset (PR) input, 96, 98, 99
prime implicant (*see* PI)
primary encoder, 65, 110–112
problem-oriented, 182
process level in computer logic, 144
product-of-sums form, 34
product operation in Boolean algebra, 32
product-term, 77, 83
program counter (PC) (*see* registers)
programmable ROM (PROM), 51, 145, 146
PROM programmer, 51
propagation delay (*see also* gates), 70, 108, 109, 131–133, 138
propagation term, 162
p-type doping, impurities, 37
pulse, 91–100, 102, 104, 130, 139, 140
pulse-triggered flip-flop, 93, 94

Q

quad gate package, 47
quartz crystal oscillator, 102, 116
Quine, W. V., 86, 189
Quine-McCluskey tabular minimisation method, 86–89

R

race, 133, 134
radiation counter, 102
random access memory (RAM), 50
random logic, 144, 149
read-only memory (*see* ROM)
reflected binary, 62
refreshing RAMs, 51

register, 19–22, 90, 99, 100
    CC, index, IR, MAR, MDR, PC, 19, 20, 152, 164–168
    implementation, 154–156
    transfers, 20–22, 165–169
resistor-transistor logic (RTL), 14
ring counter, 170
ripple-through counter, 108
ripple-through effect, 116
rising edge, 93
ROM, 50, 51, 145–147, 149
rotate instruction, 154
RTL, 14, 36

S

saturation of transistor, 38
schduler, 181
Schmitt trigger, 138, 139
Schottky-clamped TTL (S), 41, 44
secondary signal, 134
second-generation digital computer, 14, 17
semiconductor manufacturers, 16
semiconductor material
    germanium, 37
    silicon, 37
sequence counter, 169–171
sequential logic, 22, 23, 57, 65, 83
    design, 90–109
    design, steps in, 101
serial addition, 70
series 54, series 74 logic, 16, 40, 47
seven-segment LED display, 103, 120
Shannon, Claude, 30, 189
shift register, 99, 100, 127, 128, 154–156
Shockley, William, 36
sign of 2s complement number, 156–157
silicon dioxide, 42
silicon-sapphire (SOS), 42, 43
single-chip microcomputer, 149
single-input NAND, 68
single-variable hazard, 134
small-scale integration (see SSI)
software portability, 178, 179
source (of MOSFET), 42
spare gates in IC package, 85, 86
speed of logic families, 39–44
speed-power product of logic families, 41–44
Sperry-Rand Corporation, 12
Spike, 93, 133, 134
sprocket hole, 142
S–R flip-flop, 28, 90–96 (see also flip-flop)
SSI, 16, 23, 28, 36, 41–47, 52, 66, 67
stack, 180–181
standard TTL (N), 41, 44
standby mode (of dynamic RAM), 51

starting state, 106
state assignment, 109
state diagrams, 63–65, 92, 100–102, 106, 107, 124
state minimisation (state reduction), 109
state table, 63–65, 91, 101, 107
state variable, 100–106
static hazard, 133
static RAM, 51
stepping motor, 142
stored-program machine, 12
SUB instruction, 159–160
subroutine, 181
substate, 41, 42
subtraction process, 156–158
sum-of-products form, 34, 68, 77, 147–149
sum operation in Boolean algebra, 32
supply voltage, 137, 138
switching algebra, 30
switching representation of logic circuits, 58, 59
switching speed, 43
switch, transistor as, 38, 39
synchronous operations, 15
synchronous sequential logic circuit, 57, 101, 104, 106, 147, 148
System/360, 16
system software, 179

T

tabular minimisation (see Quine-McCluskey method)
tagged architecture, 182
target computer, 178
Texas Instruments, 16, 47, 193–206
theorems of Boolean algebra, 59, 60
third-generation digital computers, 15, 17
thrashing, 184
timing diagram, 92, 97
timing wave form, 130, 131
toggle, 99, 100, 108
tolerances of components, 134–137
totem-pole output, 39, 44, 135
traffic lights controller, 128–131
trailing edge, 93, 100, 139
transceiver, 152
transistor, 14, 16, 36–42
transistor-transistor logic (see TTL)
transition time, 138
transmission errors, 113
triple gate package, 47
tri-state output, 44, 143, 152, 153, 165
truth table, 27, 58, 64, 75, 146, 148
TTL, 16, 84
    family, 43, 44
    gate, output stage of, 93

TTL (*continued*)
  gate loading, 135–137
  gates, 36–44
  logic, forms of, 41–44
  logic levels, 93
  logic, the five types of, 44
  TTL–compatible, 136–138
  TTL–MOS, TTL–CMOS interfaces, 137, 138
  typical voltages, 137
Turing, A. M., 12
Turing machine, 12
twos (2s) complement number, 156–158
types in programming, 182

U

uncommitted logic array (ULA), 148, 149
unconditional branch instruction, 175
union of sets, 60
unipolar logic, 42
unipolar transistor (*see also* transistor), 36, 41
UNIVAC, 12
universal logic module, 144, 145, 149
unused inputs, 131
utility program, 9

V

V_{cc}, 45, 47, 93, 96, 135, 136
vacuum tube, 14, 83
Venn diagram, 60, 61
vertical micro-instruction, 176, 177

virtual machine, 179
virtual memory (VM), 14, 182–184
visual display unit, (VDU), 151
VLSI, 36, 185
volatile memory, 50
voltage regulator, 44
von Neumann, J., 12–15, 187
von Neumann machine, 12, 13, 18, 19, 150, 152, 184–186
VM (*see* virtual memory)

W

waveform, 130
width of pulse, 93, 94
Wilkes, M. V., 14, 173, 191
Williams, F. C., 14
wire-wrapping, 70
wired–AND, 44
wiring diagram, 45, 68–70
word, 48–50
word length, 19
word level in computer logic, 24, 144
working set model, 184
worst-case values, 136
write enable, 49
writeable control memory, writeable control store (WCS), 173

Z

Z1, Z2, Z3 computers, 12
Zuse, Konrad, 12